Achieving Global Information Networking

For a complete listing of the *Artech House Telecommunications Library*, turn to the back of this book.

Achieving Global Information Networking

Eve L. Varma
Antonio Rodríguez-Moral
George W. Newsome
Dennis K. Doherty

Thierry Stephant
Christine Pageot-Millet
Jean-Michel Cornily

Lucent Technologies

France Telecom CNET

Artech House
Boston • London

Library of Congress Cataloging-in-Publication Data
Achieving global information networking / Eve L. Varma . . . [et al.].
        p.   cm. — (Artech House telecommunications library)
    Includes bibliographical references and index.
    ISBN 0-89006-999-9 (alk. paper)
    1. Telecommunications systems—Specifications Standards.
2. Telecommunications systems—Models. I. Varma, Eve L.
II. Series.
TK5102.5.A2858 1999   99-33670
621.382—dc21   CIP

British Library Cataloguing in Publication Data
Achieving global information networking. — (Artech House
    telecommunications library)
    1. Information networks 2. Telecommunications—Computer
    networks—Design 3. Telecommunications—Computer networks—
    Standards
    I. Varma, Eve L.
    621.3'821

    ISBN 0-89006-999-9

Cover design by Lynda Fishbourne

International Standard Book Number: 0-89006-999-9
Cataloging-In-Publication: 99-33670

10 9 8 7 6 5 4 3 2 1

Contents

Foreword

It is a natural human tendency, when faced with complexity in the real world, to simplify by a process of abstraction and classification so that uncountable millions of perceivable things are reduced to a tractable number of basic entities or concepts characterized by their specific attributes and relationships. These entities become the actors in our mental model of the world and their attributes and relationships constrain the roles they can play in our model. This general tendency is confirmed in our understanding of the natural world as well as the more specialized worlds of human organizations and physical artifacts including telecommunication networks. Specialists in one area sharing a common model can communicate effectively within their specialization, but specialists from another area who do not share the same model cannot. It is possible for two worlds to communicate so long as they have a shared model whose scope is sufficient for their mutual interest, which will often be smaller than the scope of each model individually.

Thus, in the late 1980s and early 1990s, several groups of telecommunications specialists representing the SDH and ATM communities came together with IT specialists under the auspices of the ITU (then called CCITT) to devise a shared model of telecommunication transport networks that could be used as the basis for common technical developments in the areas of network management and control in a multioperator, multivendor environment. It was in pursuit of this goal that I first met many of the authors of this book. My own books written together with Andy Reid of BT on Broadband Transport Networking derived directly from this period. Although written from the telecommunications viewpoint, these also contained an account of the rationale behind the layered model that had been developed primarily to explain telecommunication transport networks to the IT specialists who would be responsible for designing software management and control systems for them.

The methodology for organizing such multidisciplinary activities was neither well developed nor understood at this time, and the formal syntax for expressing the results left much to be desired. Distributed computing was still a research topic, and the more limited hierarchical architectures inherited from the TMN were the normal targets for telecommunication system software. The present book starts from here but makes a much broader rationale for the central role this sort of modeling activity has in systems development far beyond the mere writing of interworking standards. It then goes on to introduce an expanded framework developed in the mid-1990s for dealing with distributed systems. Traditional TMN architecture based on a strictly hierarchical, master-slave model exhibited serious limitations in performance and scalability. New distributed architectures like CORBA bring a large repertoire of high performance possibilities within range.

The situation today also differs from before in that modeling tools are now available based on these principles to support the new expanded methodology for distributed systems. The designer working in this field today can express his model clearly using commercially available tools and develop it through validation and testing to code generation. The authors illustrate their explanations with real examples using the tools based on the unified modeling language (UML).

I wish Eve Varma and her team of contributors well with this book and I confidently expect that the methods described here will become the mainstay of network software system designers for several years to come.

Mike Sexton
August 1999

Preface

Specification of today's complex, interactive, and interoperable network and equipment applications presents a tremendous challenge, and has been the driving force for improvements in how network and equipment requirements are expressed. In particular, significant progress may be achieved through use of more rigorous "model-based" techniques that are better suited to meet this challenge. Using these techniques, we may also characterize the relationships between entities traditionally thought of as being in different domains (e.g., software and telecommunications, services, management, and transport). However, while such techniques are of great importance to telecommunications engineers, there are no books currently on the market that provide a cohesive treatment of them. The objective of this book is to fill that gap.

Professor B. Gopinath, then of Rutgers University, once suggested that nearly all behaviors that occur in the universe might be characterized as resulting from interactions among networked entities across various types of interfaces. From molecular to galactic structures, entities engage in relationships that, via appropriate characterization means, can be used to describe behavioral events. When we talk about the interplay of human-engineered entities, such as modules within a software system, or components on a circuit board, this principle can be used to describe the *desired* (rather than observed) behavior of systems. At some level, all behavioral specification approaches are based on this fundamental principle.

The unique observation of the authors of this text is that this same basic principle can be used to characterize the interactions that occur among the different domains of abstraction that comprise various systems, and certainly among most of those that comprise telecommunications networks. In the past, "constructionists" asserted that the nature of the universe could be understood simply by understanding the nature of its constituent atoms. They were proven wrong when it was realized that complex systems develop consequential behaviors resulting from their interaction with other complex systems, and that these behaviors cannot be explained in terms of an isolated understanding of the individual atomic components (for example, life). In other words, the whole is often greater than the simple sum of its parts. In order to have some understanding of the complex system, it is necessary to have some understanding of how the elements of the system interact. Specification of complex computer and communications systems requires an equally detailed understanding of the dynamic relationships that exist between the various domains that might otherwise be specified in isolation.

For example, a particular piece of telecommunications equipment within a network consists of entities in both the hardware and software domains of abstraction, each of which is modeled according to the interactions within their networks of components. However, to accurately and completely characterize the behavior of the system as a whole requires not only that we provide independent specifications of the hardware and software behaviors, but that we also characterize the interactions that lead entities in one of these domains to have impact on entities of the other domain. And this only defines the behavior of an individual system within a telecommunications network. In order to be able to accurately and completely specify the desired behavior of the network itself, which comprises numerous systems, we need to adequately characterize the interactions of these individual systems with each other, and at all relevant levels of abstraction. These domains of abstraction can include transport and management and all the various levels and technologies that make up the hierarchy of service and infrastructure networks.

Today we find ourselves at a time in the evolution of the telecommunications industry in which demand for improvements in integration and performance, and for interoperability of various vendors' equipment and network management systems is at an all-time high. Telecommunications network providers realize that the key differentiators in their competitive business are the services offered and the efficiency with which they manage the networks supporting those services. However, there are many factors that increase the complexity of modern networks and impact the degree of planning involved in establishing support of associated services.

This book is the first to offer a cohesive approach towards meeting the challenge of supporting complex, interactive, and interoperable network and equipment applications. It is also the first definitive publication explaining the need for, implications, and usage of key model-based approaches utilized within global standards for the specification of telecommunications networks, equipment, and their associated management.

The introductory chapters motivate the usage of model-based specification techniques by explaining the traditional role network modeling has played in achieving telecommunications network operator and manufacturer objectives, and examining requirements capture drivers and criteria based upon current and emerging customer, market, and technology considerations. Conventional network infrastructure management and specification approaches are then reviewed in sufficient detail for the reader to understand their fundamental concepts and features, and assessed in the context of the requirements criteria that have been established.

The principal part of the book is devoted towards enabling the reader to understand, apply, relate, and unify two key model-based methods that are being utilized within the global standards community. These are the functional modeling and open distributed processing (ODP) approaches, which are used to specify telecommunications networks and equipment, and telecommunications network management, respectively. Within the global community, usage of the functional modeling approach, as a *domain-specific* technique for the specification of transport functionality, is generally recognized as key to enabling unambiguous specifications and equipment interoperability. Global equipment vendors and network operators need to understand, generate, and determine compliance with such specifications. This book provides sufficient material and examples to enable the reader to achieve this level of proficiency. The ODP approach provides a generic framework for specification of information technology (IT) applications and has been used for the specification of management functionality within the telecommunications industry. This book provides the reader with a working knowledge of the application of ODP concepts to telecommunications network management, enables understanding of associated standards specifications, and provides insights into the evolution of Telecommunications Management Network (TMN) concepts. After introducing and elaborating these approaches, the book then folds specification of transport behavior into the same framework as specification of management behavior, enabling a holistic view of the overall network behavior, which is then rolled into a more generally useful framework for services.

The final segment of the book provides an overview of a key standard specification language that may be used to express ODP-based management specifications and introduces fundamental concepts and realization approaches associated with interdomain management. It culminates with provision of a multi-technology application example that involves specification of a particular service and illustrates how this specification translates into associated network and equipment specifications. We introduce the reader to the *unified modeling language* (UML) because it is rapidly becoming the de facto standard specification language, and to interdomain management because of the industry need to enable interoperability between management systems based upon different technologies and from different vendors. The multi-technology application example illustrates the usage of, and interrelationships among, all of the concepts and models introduced in the preceding chapters.

We would like to express our thanks to those who helped us in the preparation and review of material for this book. In particular, we are indebted to Mike Sexton, of Alcatel, for his critical review of the entire draft and for his insights into transport network architecture and functional modeling. We also thank Eric Debeau, Jean-Marc Pageot, and Maarten Vissers for their review comments and sharing their expertise. Our sections on TINA-C would not have been possible without Fabrice Dupuy, and our understanding of ODP was greatly enhanced by Pramila Mullan and Tom Rutt. We are similarly indebted to Patrick Juré, Patrice Lamy, Jean Lawlis, John Ellson, John Islip, and John Bunting for their insights into network modeling and management, and to J. Barrie Leigh, of Fothergill Ltd., for stimulating our thinking on systems engineering and presentation techniques. The authors would also like to acknowledge all the experts involved in related standards activities, with whom we have had many stimulating sessions and from whom we have learned a great deal, and especially Malcolm Betts and Chuck Widowson of Nortel, Andy Walsh of Telcordia, and Bernd Zeuner of Deutsche Telekom. Last, but by no means least, we would like to thank our management and families for their tolerance of this endeavor, which, like so many network management projects before it, ended up being far more complex than envisaged.

1

Role and Impact of Network Modeling

1.1 Telecommunications Network Vision

It is becoming clear to telecommunications network providers that the key differentiators in their competitive business are the services offered and the efficiency with which they manage the networks supporting those services. Many significant challenges remain to be met as new services are launched and as the technology and topology of networks change. Next generation network architectures for cost-effective, reliable, and scalable evolution will employ both transport networking and enhanced service layers, working together in a complementary and interoperable fashion. This next generation network will dramatically increase, and maximally share, backbone network infrastructure capacity and provide sophisticated service differentiation for emerging data applications [1]. In the context of such network evolution, there is an increasing drive to offer customers new services, to reduce charges, to improve quality, and to offer customer control of service. Crucial to achieving rapid and flexible service provisioning is the need to integrate network administration functions fully with, for example, service management functions [2].

This has never been more true than it is today, where public telecommunications are increasingly relying upon an extremely diverse network of networks, with widely varying topologies, deployed technologies, services, and underlying business models. As is illustrated in Figure 1.1, networks are migrating toward the integrated use of multiple technologies organized into "layers" with client-server relationships existing between layers. The example illustrates possible layering scenarios involving a range of networking technologies; that is, plesiochronous digital hierarchy (PDH) networks (e.g., 1.544 Mbps and 2.048 Mbps), packet-switched networks (e.g., asynchronous transfer mode, or ATM, and

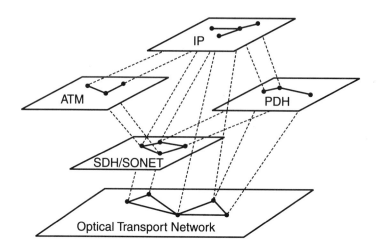

Figure 1.1 Multilayer networks.

internet protocol, or IP), synchronous digital transport networks (e.g., SDH, or synchronous digital hierarchy, and SONET, or synchronous optical network), and optical transport networks (OTN). Operators provide high-quality services by exploiting the client/server relationships between these network layers. For example, we may want to provide additional capacity via control of an optical transport network in support of a request from an SDH network that, in turn, is being asked to provide for the transport of new IP services. Complexity is introduced because technologies are mixed based on perceived optimality to support particular services and other operator-specific criteria such as cost and performance. This leads to an arbitrary variety of possible combinations to support.

During their whole life cycle, networks and services must be managed, encompassing such tasks as forecasting, service and network requirements analysis, detailed dimensioning, configuration data management, inventory management, maintenance and repair, traffic management, capacity assignment, and performance analysis. The challenge is to encapsulate all of these functions so that they may effectively interact and be applicable to all networks. An operator vision can embody end-to-end network management, coverage of the whole network life cycle, fully integrated functionality, flexibility (evolution and growth), high levels of automation/decision support, and conformance to architectural objectives [2]. From a vendor perspective, the challenge is to cost-effectively provide products that support this vision. The question for network providers and vendors alike is how this may best be accomplished. This book will demonstrate that using network modeling techniques as a means to establish a successful set of integrated requirements will enable realization of this vision.

This book intends to concentrate on the transport and management domains, describing others where it assists comprehension and exploring the relationship with the control domain. We will demonstrate that the model-based techniques we will describe herein are applicable to all problem domains exhibiting analogous characteristics, including packet-switched and wavelength division multiplexed (WDM) networks, and integrated combinations thereof, to support a variety of service scenarios.

1.2 Background

In a broad sense, a network is a system of interconnected elements. Topologically, it can be represented by a set of nodes and a set of links that interconnect pairs of nodes. A network is needed when communications services must be provided to many, widely dispersed customers. Depending on the types of services, the characteristics of the network elements may differ greatly. Another concept necessary for the definition of a telecommunications network is the notion of telecommunications traffic. Traffic is the flow of information or messages through the network and may consist of simple telephone conversations, or it may be the result of complex data, video, and audio services. Thus, a telecommunications network is a system of interconnected transport facilities (i.e., transport network infrastructure) designed to carry the traffic that results from a variety of telecommunications services [3]. Many factors determine the structure of the transport network. Some of the more important factors are customer location; telecommunications services desired; service access point locations; service characteristics including performance objectives, available communications technologies, and their costs and performance characteristics; and network survivability features to protect against major service interruptions.

Network planning is a multifaceted discipline that encompasses the functions involved in planning the evolution and implementation of the network, designing and engineering the configuration of the network, and managing the total network investment. Its objectives are to provide network capacity and to guide the operations functions in using this capacity to meet customer demands in a cost-effective manner. One result of network planning is the establishment of plans for various aspects of the organization and operation of the network. For example, network configuration planning begins with a forecast of demand for various services; currently available and projected technology; the existing configuration of the network; and regulatory, legal, structural, financial, and other constraints. Its specific goal is to determine the future configuration of the transport network as a function of time so as to meet the increased demand for a variety of services in an orderly fashion while minimizing total network costs. In addition, transport networks and equipment must be properly installed and maintained in order to provide telecommunications services that meet quality-of-service (QoS) objectives. Thus, network planning is not a process of performing various tasks in separate, isolated planning disciplines and then combining the results into a single plan. Rather, it is a continuous process within each of the disciplines with significant interactions among them so that changes in one may affect plans in others [3].

To design networks of such complexity as current and future telecommunications and data networks, it is imperative to describe them first in terms of models. Models are the result of abstracting some portion of reality by simplification through the employment of a defined set of modeling constructs. Modeling networks allows us to overcome their inherent complexity, concentrating on the aspects relevant to some specific problem, such as that of specifying the structure or the behavior of the whole network or some portion thereof. Modeling techniques vary in their degree of formalism and rigor. The purpose of any modeling technique, regardless of its notation and syntax, is to correctly convey semantics. Today the popularity of using formal modeling techniques is increasing in almost all application areas due to the widespread availability of computing capabilities that ease the tasks of information compilation, syntactic model checking, and model simulation and verification. Each specific knowledge domain, from electronic hardware design to artificial neural networks, has developed application-specific syntax and notation.

Virtually every component of a network planning study has to be a model, or approximation, of the real situation. Obvious examples of this are the network topology, the physical and cost characteristics of the available technologies, the various types of traffic demands, and different engineering constraints. The degree to which network planning studies can utilize model-based approaches depends upon the limitations or capabilities of the modeling techniques employed. For example, if the various components of the problem are described in very detailed, realistic terms, the sheer size of the problem may render it intractable. On the other hand, if the problem is represented by highly aggregated models, it becomes difficult to accurately translate the solution for the modeled problem into a solution for the real problem. Thus, the planner must determine which components of the problem should be modeled at which level of detail. Often, the only guidelines that have been available have been the planner's experience and the extent of his or her knowledge of the network [3]. However, as the complexity of the network increases, advanced technologies are deployed, and the demand for many complex telecommunications services increases, traditional planning methods become increasingly suboptimal.

Network planners and equipment developers must translate their proposed solutions into sets of requirements. Some characteristics of successful requirements are that they are complete yet minimal, unambiguous, consistent, traceable, verifiable, and correct. Requirements that are complete yet minimal means that only necessary and sufficient aspects of the solution are specified. Overspecification can unnecessarily constrain design and development and reduce innovation. Unambiguous requirements are clear and not subject to alternative interpretations. Being consistent means that various aspects of the requirements are feasible, not only individually but also together. Traceability involves the ability to trace the sources and impact of any individual requirement. Verifiability is the capacity to determine that the requirement has been met. Correct is, of course, the most important characteristic. It means that the requirements actually reflect a "right" solution to the "right" problem. Generating a cohesive set of requirements that spans the services, network, and equipment domains to support the network service and equipment provider vision is a significant challenge to be met.

1.3 Linkages Between Transport Services, Networks, and Equipment

Transport networks are designed by network architects to support the requirements of the services they carry. To perform this task, they have to take into account the following inputs:

- Location of transport service access points;
- Traffic demand pattern;
- Transport technologies and characteristics such as signal rate and format;
- Expected quality of transport service;
- Available equipment;
- Cost of the possible solutions;
- Infrastructure constraints;
- Organizational constraints.

Application planning helps architects to integrate all of the aforementioned parameters, with the output being a network architecture expressed in terms of topological indications. Due to the complexity of the problem and the range of potential solutions (which are not all equally feasible), model-based simulation tools would significantly aid gaining closure on what constitutes the best architecture.

The output of the network planning phase leads into the equipment supply and provisioning phase where, in parallel, network databases are elaborated. The purpose of a network database is to logically model the topology of the real network. In network databases, access points are created depending on the provisioned equipment (e.g., "tributary unit" on a cross-connect system). In the same way, subnetworks will be associated with matrix capabilities and, depending on client policies, organized into composite subnetworks. This composition produces network abstractions that will hide the detail of actual realizations and, consequently, the detail of operations for network clients. Obviously, the provisioning phase will be constrained by the planning phase, and created resources in the network databases are in fact resources that move from a planned state to a commissioned state.

The real traffic, which will reflect the number of transport service requests, need not necessarily correspond to the provisioned traffic forecasted during the planning phase. For example, what if the client locations are not those that were planned, or what if the traffic demand is increasing faster than anticipated? To ensure that sufficient network capabilities are available to handle actual demand, the network utilization needs to be periodically compared to the provisioning forecasts; if these are found to be misaligned, a replanning of the network could be required. Due to the fact that infrastructure investments (e.g., equipment, cable, and operation systems) need to be optimized with respect to traffic levels, and due to the inability to predict transport demand accurately, it will become more important to be able to revisit the planning process frequently and to adjust (as closely as possible) the network to support the actual traffic demands. Thus, planning phases, provisioning phases, and real-time operations need to be strongly coupled. This will be possible if the associated models are consistent and easily related.

A network is optimized when:

- It is able to satisfy any transport request while minimizing the amount of idle resources.
- Each involved transport entity uses the minimal amount of active resources.

These criteria need to be taken into account during all the network design phases—from the planning activities to the equipment selection, and for real-time operations. Network optimization involves a recursive application of all the previous phases. This is a new characteristic, as compared to the past where these phases were not so strongly coupled, leading more to a continuous process starting from the service requirements, passing to the network design, the infrastructure allocation, and the real-time operations. Such an overall process will be strengthened by the existence of a unified model.

A telecommunications service must at least include a function responsible for the transport of information. This function is performed by the invocation of transport services. A transport service represents a contract established between a transport service caller and a transport service provider to transport caller information across the network provider domain. The caller and the provider roles can be fulfilled by administrative entities such as human resources or companies. A transport service provider has to select and bind a set of transport resources to fulfill the transport service requirements. The transport network infrastructure is inherently comprised of heterogeneous hardware, software, and communications protocols and is often constrained by organizational considerations as well.

The selection of resources will be made depending on the characteristics of the information that is to be transported across the network domain and the quality of the transport service that is negotiated between the caller and the provider. The selected resources may include transmission equipment, switches, cables, databases, and routing applications, for example. Together these resources will form a telecommunications network infrastructure. The selection of the needed resources by the transport provider is tightly coupled with the transport characteristics requested by the caller, such as transport technology (rate, format), availability, performance, and cost. In that sense, telecommunications services will drive the choice of transport technologies. On the other hand, the selection of services offered will depend on the available resources from the provider side. Thus, the availability of telecommunications services will be conditional upon the available technologies and the associated equipment. It is not of great importance to know if the technology is the driver of the service or vice versa. A transport service provider has to take into account the technologies available in the industry when it constitutes its offer, while telecommunications equipment providers have to design technologies anticipating the potential for telecommunications services.

In summary, it is necessary to maximize the synergies between the transport network infrastructure and the desired catalog of services while optimizing solutions for telecommunications service providers in terms of characteristics/cost/performance metrics and developing a consistent solution in a form that masks the heterogeneity of the infrastructure components.

As described earlier, network planning requires the use of modeling; however, the value it brings depends upon the quality and appropriateness of the modeling approach employed. This book will demonstrate that an ideal model has to consider all of the involved characteristics, ranging from the end-users' perception of a service to the underlying transport network infrastructure.

In terms of facilitating the generation of requirements, it must provide a specification technique that enables a unified description of the behaviors of services, networks, and equipment. This technique must enable these behaviors to be specified in a totally functional manner, independent of any notion of how they might be physically distributed in a solution system or what technologies might be used to build such a system.

If service, network, and equipment behaviors were to be completely specified in terms of interfaces between functional entities, it would be possible to exploit different physical implementations, exposing various instances of these interfaces in support of optimization of cost, performance, and other factors. Obviously, such a model would not be visible to, or known by, all parties involved in offering telecommunications services solutions. For example, the equipment designer is not directly concerned by service management considerations, while the telecommunications service specifiers do not concern themselves with the details of equipment specifications. However, each component has a role to play that is consistent with its environment, leading to the existence of a complete solution, which does not assume any partition between transport characteristics and service characteristics, and that is distributed among all of the involved parties.

1.4   Linkages Between Transport and Management

Transport behavior and management behavior have traditionally been specified in separate organizations, have employed different specification methodologies, and have no clear "mapping" to facilitate understanding of the relationships between them even though dependencies clearly exist. Including transport behavior in management specifications enables the creation of unified specifications that maximize interoperability and reuse while clarifying the relationships that exist between all facets of a system's or network's behavior.

Telecommunications services and transport network management have traditionally also been specified in different organizations even though service behavior is also a part of the overall network behavior that must be understood in order to facilitate interoperability. While it is possible to gain some benefit from the complete specification of transport network management behavior, it is clear that additional benefit can be gained from exploiting the commonalities between service and management behaviors since both are essentially manipulations of network resources. The two sets cannot be developed independently, even where they are constituted of overlapping subsets, due to the recursive nature of the relationship between them (e.g., intelligent network features). However, transport services are assumed to be generic and act as communications support features for the more general telecommunications services. Consistency between these sets of specifications is essential to ensure a thorough understanding of what the management functions and network resources deliver and expect from each other.

1.5 Unification of Transport and Management Modeling Approaches

From the perspective of the client of any telecommunications network, the distinctions between the management and transport behaviors of the network are less important than the behavior of the network as a single integrated entity. In fact, both of these kinds of behavior can be thought of as describing what happens when the network is in various states. For example, mechanisms to describe transport behavior typically characterize part of the behavior of the network as it occurs in a steady or stable state. Management operations most often transpire in a transient state, such as when connectivity is being established or when a connectivity or quality-related problem has occurred. A holistic view requires that we understand the relationships between the transport and management domains.

There are several reasons why this issue has largely been ignored in the past. Because of the nature of the technologies used to implement these behaviors, it had been fairly easy to separate our means of describing them. Transport behavioral descriptions were used to characterize the intended design or function of the network in delivering its service, normally thought to be fairly stable once established. Management behavior descriptions characterized the mechanisms put into place to establish the connectivity or to report observed problems. Consequently, network service providers typically organized themselves to align with these functional boundaries.

More recently, these boundaries have become blurred. Automated provisioning has become the result of a cooperative and synergistic relationship between digital cross-connect systems, or other systems containing configurable connectivity matrices, and the network management systems. Reconfiguration activity for service survivability may be embodied entirely within the network elements, with no real-time support from any management system.

As the systems that embody management and transport functions become increasingly amorphous and indistinguishable from each other, considerable debate and confusion about the definitions of these functions has arisen. It has become increasingly important to understand the behavior of the network as a whole, regardless of categorization into transport or management behavior (although, as will be seen later, there are still times when the "separation of concerns" provided by this distinction is still valuable). Thus, we observe that advances in networking technologies have led us, even at the network element level, to adopt the holistic view of the network client.

1.5.1 High-Level Description and General Principles

Thus, we highlight the need to provide a unified model for transport and management that describes network capabilities, network management capabilities, and relationships with equipment and equipment management. Earlier modeling approaches have not generally accommodated this need. However, the principles behind a unified approach are now well understood, and more recent modeling approaches can facilitate its realization.

For example, within global standards, an extensive set of specifications has been generated relating to synchronous digital equipment [4] and its management [5–12]. Much effort has been devoted to aligning equipment functionality pertinent to its management and the associated management information model. As will be described in Chapter 6, it is possible to map between the functional specifications and management specifications. However, basing both equipment and management specifications upon a common model assures consistency in accordance with a holistic view of network behavior as described previously.

A major goal of the specification of network behavior is the clear and complete description of interfaces sufficient to provide interoperability. As we will describe more fully later, a unified model for describing all entities that play a role in a network must address three major needs to provide specifications sufficient to ensure interoperability:

- Unification of transport and management behaviors to ensure consistency;
- Formal, precise semantics of object behavior to prevent ambiguities;
- Detailed, object-oriented, distribution- and technology-independent specification of network-level management applications.

Established equipment management standards are not sufficient, in themselves, to provide equipment-level interoperability because only the internal behavior related to management and some part of the cooperative behavior with the internal transport behavior can be described with these management standards (Figure 1.2).

However, we will demonstrate that it is possible to integrate both the transport and management behaviors from the relevant standards into a unified model of the entire relevant equipment behavior (Figure 1.3).

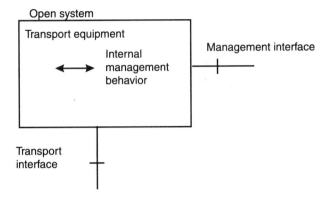

Figure 1.2 Equipment interfaces and management behavior.

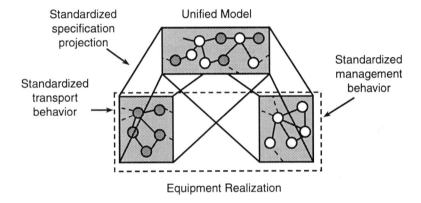

Figure 1.3 Unified model of equipment behavior.

The previous discussion addresses behavior at an equipment level. Still other standards have been created to describe the network-level view of management applications that manipulate the element-level resources or manipulate atomic functions that are used to compose network transport services. We can also illustrate the mapping between a transport subnetwork and its associated network-level transport functional model and management model implementations, as illustrated in Figure 1.4.

Similar to how we described the integration of equipment-level management and transport behaviors, we can provide a model that unifies the network-level management and transport behaviors as well. These standardized realizations could be reflected in a unified transport and management network-level model as illustrated in Figure 1.5.

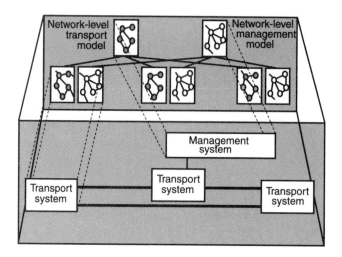

Figure 1.4 Relationship between network- and equipment-level transport and management models and their equipment realizations.

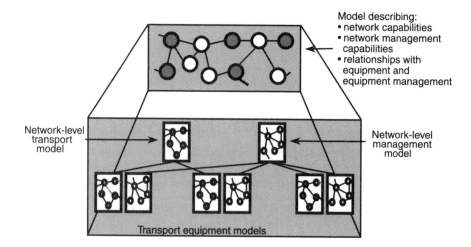

Figure 1.5 Unified model of network-level behavior.

1.5.2 Applicability to Various Transport Technologies

The unified view described in Section 1.5.1 is inherently technology-independent. Since the behavior of the network is modeled functionally and holistically, the model abstracts away the mechanisms that would ultimately implement this behavior. Thus, the approaches described in this text can be applied to any network satisfying the characteristics we described earlier: a collection of nodes and links serving the common purpose of transferring client service among disjoint locations. Due to the technology-independence of the approach, it becomes possible to capture the client-server interactions between, for example, an IP network and an underlying transport network (as described in Section 1.1).

1.6 Summary

The objective of this book is to describe the use of model-based approaches to achieve network planning, architecture, and interoperability goals. There are a number of criteria that an approach must satisfy in order to achieve these objectives. Among these criteria is the need to present the behavior in a holistic, unified manner, an area in which most approaches to date have been found lacking. This integration is driven in part by the erosion of a correspondence between management and transport functions and the mechanisms used to support them and also by the trend toward convergence of transport and data networking.

It is important to recognize that, as a pragmatic consideration, modeling approaches must evolve more or less incrementally toward a more complete satisfaction of their objectives. This is due to the embedded base of specifications using earlier approaches, which, if overturned in a revolutionary rather than evolutionary manner, would promote much of the chaos such new approaches are designed to prevent.

References

[1] Alferness, R., P. Bonenfant, C. Newton, K. Sparks, E. Varma, "A Practical Vision for Optical Transport Networking," *Bell Labs Technical Journal,* Vol. 4, No. 1, Jan.–Mar. 1999.

[2] Helleur, R. J., and N. R. P. Milway, "Network Management Systems: Introductory Overview," *British Telecommunications Engineering,* Vol. 10, Oct. 1991.

[3] AT&T Bell Laboratories, "Engineering and Operations in the Bell System," Second Edition, 1984.

[4] ITU-T Recommendation G.783, "Characteristics of Synchronous Digital Hierarchy (SDH) Equipment Functional Blocks," published version Apr. 1997, draft revised version, Feb. 1998.

[5] ITU-T Recommendation G.774 "Synchronous Digital Hierarchy (SDH) Management Information View for the Network Element View," published version Sept. 1992, corrigendum 1, Nov. 1996.

[6]   ITU-T Recommendation G.774-01 "Synchronous Digital Hierarchy (SDH) Performance Monitoring for the Network Element View," published version Nov. 1994, corrigendum 1, Nov. 1996.

[7]   ITU-T Recommendation G.774-02 "Synchronous Digital Hierarchy (SDH) Configuration of the Payload Structure for the Network Element View," published version, Nov. 1994, corrigendum 1, Nov. 1996.

[8]   ITU-T Recommendation G.774-03 "Synchronous Digital Hierarchy (SDH) Management of Multiplex Section Protection for the Network Element View," published version, March 1996, corrigendum 1, Nov. 1996.

[9]   ITU-T Recommendation G.774-04 "Synchronous Digital Hierarchy (SDH) Management of the Subnetwork Connection Protection for the Network Element View," published version, July 1995, corrigendum 1, Nov. 1996.

[10]   ITU-T Recommendation G.774-05 "Synchronous Digital Hierarchy (SDH) Management of the Connection Supervision Functionality (HCS/LCS) for the Network Element View," published version, July 1995, corrigendum 1, Nov. 1996.

[11]   G.774-06 "Synchronous Digital Hierarchy (SDH), Unidirectional Performance Monitoring for the Network Element View," published version, April 1997.

[12]   G.774-07 "Synchronous Digital Hierarchy (SDH) Management of Lower Order Path Trace and Interface Labeling for the Network Element View," published version, Nov. 1996.

2

Transport Network Infrastructure and Management—Specification Approaches

2.1 Introduction

In this chapter, we discuss some common approaches to specifying network infrastructure and its management. We begin by reviewing requirements capture drivers and criteria, including customer or market needs and technology factors, and establishing some criteria for successful requirements specifications, and we then proceed to describe current approaches, many of which are being broadly used within the telecommunications industry. We conclude with a characterization of the degree to which these approaches appear to satisfy requirements capture criteria and use this knowledge as a foundation for the succeeding chapters.

2.2 Requirements Capture Drivers and Criteria

Requirements definition processes are intended to provide an orderly and iterative definition of a problem and development of a solution. Much work has been devoted to establishing processes for requirement analysis and definition. Typically, the de facto approach used for network and equipment requirements definition involves a number of phases of specification, generally followed in sequence, until a detailed set of requirements sufficient to begin implementation is established. These may be considered as levels of abstraction associated with problem domain definition and solution domain specifications. An underlying assumption is that requirements can be cleanly "handed-off" from one phase to the next—in other words, that the levels of abstraction are "enforceable." This can be particularly important in the case where different organizations are held responsible for the various phases of specification. These phases can be generally characterized as follows:

1. Generation of a very high level specification of the problem domain such as customer or market needs in a particular area;

2. A refined specification expressed in terms of desired features and capabilities;

3. Generation of more detailed requirements that support the same customer need but do not constrain the implementation;

4. Establishment of an architecture specification that is sufficiently specific to begin implementation.

It is always an objective to establish traceability between requirements, functions, and system elements. In order to establish whether any particular requirements capture approach is sufficient to meet the needs of the particular problem domain, it is important to understand the contextual scope in addition to relevant enabling technology. Within the next sections, we discuss these aspects and attempt to characterize associated criteria that can be used to establish the sufficiency of various requirements capture approaches.

2.2.1 Customer/Market Drivers and Implications

There is increasing demand within the telecommunications industry for improvements in performance and for interoperability of various vendors' equipment and network management systems. Operators are interested in assuring that the transport and management architectures deployed within their networks provide high levels of performance and are optimized to meet their unique network needs. Many operators are also advising vendors that they would like to see network management applications from multiple vendors working together (that can be resident within the same computing platform or on different computing platforms) as well as the interoperation of various vendors' transmission equipment.

The effort required to manage a modern telecommunications network, even separating the domains of the network that have been designated as "transmission" and "switching," has grown beyond the capabilities of any single computing platform. Thus, in order to achieve the preceding objectives, it is necessary to consider the cooperation of a multitude of computing platforms. It is also clear that we cannot hope to arrive at a single physical management architecture that will provide optimal performance for each operator. The architecture of the distribution of the software among computing platforms will be as much a means for satisfying performance objectives as the architecture of the software within any of these platforms. This implies that required specified interfaces between management functions may be either internal or external to any given computing platform. It should be noted that similar considerations hold for the design of the transport network. In this case, as well, individual operator needs may vary in terms of the types of services offered and associated availability and QoS guarantees. Here too, we cannot hope to arrive at a single physical transport architecture that will optimally meet the unique needs of each operator. Thus, the ability to flexibly package transport functionality into different types of equipment supporting a range of architectures will be a key characteristic of evolving network management solutions. This also implies that interfaces between transport functions may be either internal or external to any transport equipment.

We also observe a trend toward organizational centralization of work centers that is, in fact, enabled by an operational trend toward more automated local control. This level of automation frees staff to work more efficiently, on a "remote control" basis, from centralized work centers. To achieve this demands the flexible distribution of processing and data because:

- The state of the network is most accurately reflected in the network equipment itself.
- Distributed databases tend by their nature to be more reliable than centralized databases.
- Performance is better when the place where decisions are made is as close as possible to where they are carried out. This is particularly true with active operations concerned with the configuration or reconfiguration of the network, such as provisioning and restoration.
- Taking advantage of the computing capabilities at the local level provides an opportunity to accommodate growth more readily than does concentration of data and computing capabilities at the work center. This is because processing capability increases more or less linearly with the increase in the size of the network.

Finally, from a network operator perspective, one of the major responsibilities is to provide transport functions that are contracted to their clients as transport services as well as to offer services that deliver "information content" (e.g., directory services). The management of these transport and information content services is often also offered to clients in the form of management services. In the past, transport and information content services and management services were developed independently. With the number and type of services being offered to clients rapidly increasing, significant client effort is being expended in their management, often involving different ways of accessing the services (e.g., media, identification, and procedures). Thus, there is an increasing need to establish similar means for accessing and managing the range of services offered. This implies a much tighter linkage between services, the supporting transport network infrastructure, and its associated management (as described in Chapter 1).

2.2.2 Technology Enablers

More and more elements of the telecommunications industry are coming to recognize that telecommunications networks can be thought of as a set of computing platforms, many of which happen to be associated with transmission function-related hardware. This subsection introduces two key software paradigms that are having a significant influence on the development and flexibility of specifications: the *object-oriented* and *distributed computing* paradigms. Because of their major impact on the data networking and telecommunications specification domains, they are covered in sufficient detail to provide a foundation for the material that follows in this and succeeding chapters.

2.2.2.1 Object-Oriented Paradigm

While relatively new, the object-oriented paradigm[1] has become quite pervasive within the telecommunications and data networking industries. In this section, we introduce the basic concepts and vocabulary associated with the object-oriented paradigm. While this paradigm is most relevant to the network management, and thus, software-related material in this book, we assume little or no familiarity with software concepts herein.

Introduction

According to Thomas Kuhn (in "The Structure of the Scientific Revolution"), a paradigm is "a set of theories, standards, and methods that together represent a way of organizing knowledge." The paradigm we use directly influences the way we view and, thus, model reality. Over time, software developers have employed a variety of paradigms to organize and manage knowledge related to their product design, encompassing the following views: functional, data/entity (captured in entity/relationship diagrams), behavioral (captured in state transition diagrams), and control (best captured by rule-based systems).

---

1. Note that the material in this section is gratefully adapted from our friend and colleague Richard C. Lee's book on object-oriented development [1].

Each of the earlier paradigms support modeling reality from a single, very specific viewpoint, but none of them reflect the way that people interact with the real world. The functional view, which provides an excellent means of thinking about how a computer goes about its job of fetching data and storing computed results, has been called the pigeonhole model of computation. However good a model it may be of computers, it does not reflect how people go about solving problems. In fact, while each of the other views listed focuses upon a different aspect of the real world, all of them employ a pigeonhole view of computation and are weak in modeling alternate views of the system under consideration.

The problem-solving view of object-oriented analysis is very different from the pigeonhole view of computing. In the object-oriented view, we do not refer to computing at all but rather create a universe of well-behaved objects that cooperate with each other via messages to perform services. We consider object-oriented analysis to use a simulation, versus pigeonhole, model of computation. This model is far more closely aligned to the problem-solving methods people use, and as such, serves as a better metaphor. Specifically, it allows us to think in terms of services that are available, how to find them, and how to provide them.

Basic Concepts

Clearly the most fundamental concept in the object-oriented paradigm is that of an object. We will define an object as a "thing," often implemented by software, that contains attributes (data) and methods (often software code) that act on the attributes. This defines the first principle of the object-oriented paradigm—*encapsulation*; that is, the object contains both the data and the methods that manipulate and access that data. Access to the data is only possible via the object services. The idea of providing access via services defines the second principle of the object-oriented paradigm—*information hiding*; that is, other objects have neither access to, nor knowledge of, the actual data used within the object or how a particular service is provided. However, we should not assume that encapsulation is merely a mechanism for providing information hiding because the two concepts are quite distinct. For example, bad design can easily create strong encapsulation that does not effectively hide information, and information can be hidden without the particular packaging of data and methods that is specific to the object-oriented paradigm.

In order for an object to use the services of another object, an interface must be defined. This interface definition[2] has two parts: a name for the service together with the arguments (data) needed to perform it. This should not be

---

2. In some programming languages, the interface definition is often called the prototype for the interface.

interpreted as implying that an object has a single interface providing a single service; in fact, an interface can provide more than one service, and an object can support more than one interface. An object gains access to the services of another object by sending a message, which conforms to the interface definition of the desired service, to a specific interface of a specific object. This defines the third principle of the object-oriented paradigm—*message passing.* What makes message passing of considerable interest is that the selection of the interface providing the service can be done very late, even as late as while the system is running. This strongly contrasts with traditional programming methods where the function that will provide a result is selected during the design of the system.

This discussion introduces the fourth principle of the object-oriented paradigm—*late binding,* which allows us to select the provider of a service at the time the service needs to be provided and to change the provider as needed. While perhaps difficult to see in terms of computer programming constructs, it is an exact model of human problem-solving behavior. How often do we choose the restaurant that will provide us with a meal just before going out to dinner?

While the preceding principles make it possible to decouple how objects interact from how objects provide their services, we still need some concepts to help us organize them. Recognizing that categorizing allows us to organize the complex world in which we live, we see that it is possible to make certain assumptions about an object that is in a particular category. For example, if an object is an instance of a category (class), it will fit the general pattern for that category and will have an expected behavior. This leads to the fifth principle of the object-oriented paradigm—*class/instance/object.* All objects are instances of a class, and instances can be created and destroyed as needed. Further, all objects of the same class provide the same services. The interfaces that define those services apply to every object that is an instance of that class. In the case of software-implemented objects, "how" the service is provided is also part of the class specification.

Not only do we organize our objects into classes, we also arrange our classes into a hierarchy from the general to the specialized. This leads us to the sixth principle of the object-oriented paradigm—*generalization.* The aforementioned hierarchy is called an *inheritance hierarchy* because more specialized child nodes are considered to inherit the properties of their more general parent class(es). Properties here relate to the services, methods, and attributes of the parent class(es) (which in turn can inherit from its own parent(s), and so forth). It is this inherited behavior that allows us to confidently expect specialized objects to behave in the same way as their parents, which is called *behavioral compatibility.* In many texts this is also referred to as *polymorphism,* and less frequently as *allomorphism*. Although generalization is a powerful concept, some relationships between objects cannot be captured using this concept. In particular, collaborations between objects to provide a service to a client are most certainly not generalizations. This leads us to the seventh principle of the object-oriented paradigm—*relationships*, which allows us to define the way classes are related to each other and how they may collaborate to provide a service.

In summary, the object-oriented paradigm encompasses the following principles.

- *Encapsulation* ensures that the barrier between the "inside" and "outside" of the object cannot be penetrated.

- *Information hiding* ensures that users of services have no idea about how the service is provided.

- *Message passing and late binding* together ensure that the provider of a service can be selected as late as possible and can be changed after the initial selection.

- *Class/instance/object* organizes objects into classes, describing the object services and instances that perform the object services.

- *Generalization* further organizes classes from the general to the specific and makes it possible to assume a certain amount of behavior from previous experience with the more general object.

- *Relationships* make it possible to describe the ways objects collaborate to perform a service.

2.2.2.2 Distributed Computing Paradigm

In this section, we provide an overview of the basic concepts and terminology associated with distributed computing. We introduce the set of challenges faced by designers of distributed systems and the OMG CORBA approach to distributed object computing.

Introduction

Computing today cannot be understood without considering distributed computing. The number of computers that remain isolated reduces every year, while the grade of connectivity of computers in diverse forms that range from small local area networks (LANs) to the Global Internet has undergone an explosive growth. At the same time, complex telecommunications systems may be composed of dozens or hundreds of hosts linked by communications networks, where each host or network element contains several processors interconnected by internal networks. Distributed computing is also recognized in the form of parallel algorithms and parallel systems, which exploit the availability of computing power and the possibility of decomposing processing-intensive algorithms into smaller pieces that can be solved simultaneously.

The lemma—*the network is the computer*—is a reality today, raising new opportunities and challenges for network operators and networking and application developers. Distributed computing technology is becoming more pervasive and robust, and its applicability to the design and deployment of telecommunications networks and services is considered of extreme importance. In fact, the distributed computing standards community is now being driven by telecommunications vendors to also address telecommunications network needs such as real-time performance, scalability, and fault tolerance.

Basic Concepts

Several commonly used terms related to distributed computing include distributed processing, distributed systems, and distributed applications. Within open distributed processing standards [2], distributed processing is defined as "information processing in which discrete components may be located in different places and where communication between components may suffer delay or may fail." Tanenbaum [3] provides the following definition: "A distributed system is a collection of independent computers that appears to the users of the system as a single computer." This definition addresses both the hardware and software aspects of distributed systems and emphasizes transparency as the key characteristic of distributed computing.

Although distributed computing means many things to many people (e.g., the interconnection of computers, simple remote processing, and the process of splitting tasks across multiple systems interconnected by a network), we will use it to refer to the general concept of computer systems whose parts are separated by space and "time." While this definition encompasses traditional multiprocessor systems, it also includes systems where the interactions between their parts have not been foreseen or designed a priori. Rather, they occur as the result of some of these parts (which are usually independent computing systems) joining the distributed system during a time span that can be limited. This dynamic is a very common occurrence in wide area or global networks of computers, such as the Internet.

There are several problems of architecture, specification, verification, organization, testing, and development that are specific to distributed systems. ITU-T Recommendation X.901 [4] on open distributed processing provides a very good overview of these factors.

- *Remoteness:* Components of a distributed system may be physically spread across space, so interactions may be either local or remote. The ability to support applications that can run locally or remotely from some other applications without needing redesign is a major driver for transparency in distributed systems.

- *Concurrency:* Any component of a distributed system can execute in parallel with any other component(s). This leads to a need for mechanisms to handle resource allocation and reliable group communications.

- *Lack of global state:* The global state of a distributed system cannot be precisely determined. Thus, applications that are designed to exploit some aspect of the global state of a system (a common occurrence) will not work in distributed environments.

- *Partial failures:* Any component of a distributed system may fail independently of any other component(s). This leads to a need for failure models that are more sophisticated than that of "simple total crash followed by total recovery" as well as mechanisms for dealing with distributed consensus, checkpointing, and recovery.

- *Asynchrony:* In a distributed system, we cannot expect communications and processing activities to be driven by a single global clock. Thus, related changes in a distributed system cannot be assumed to take place at a single instant. The lack of a common time framework leads to the need for clock synchronization schemes or schemes that rely on logical time. In this case, mechanisms for consistent snapshots are required.

- *Heterogeneity:* There is no guarantee that components of a distributed system will be developed using the same technologies, and the set of technologies utilized will certainly change over time in any event. This heterogeneity may occur with respect to hardware, operating systems, communication networks and protocols, programming languages, and applications. Nevertheless, means need to be established to support seamless operation of such a distributed system. Shielding developers and operators from the complexity this heterogeneity introduces is the main reason for providing transparencies in distributed systems.

- *Autonomy:* A distributed system can be spread over a number of autonomous management or control authorities (administrative domains) with no single point of control. The degree of autonomy specifies the extent to which processing resources and associated devices (e.g., printers, storage devices, graphical displays, and audio devices) are under the control of separate organizational entities. This leads to the need for addressing interdomain management issues (further discussed in Chapter 8).

- *Evolution:* Over its life cycle, we expect a distributed system to undergo changes motivated by a variety of factors. These may include technical progress that enables better system performance at a lower cost, strategic decisions regarding business goals, and the introduction of new types of applications.

- *Mobility:* The physical location of information sources, processing nodes, and users may change over time. Programs and data may also be moved between nodes, for example, to optimize performance or in order to cope with physical mobility from applications such as mobile telephony. The distributed system must continue to function under such conditions of mobility.

Various technical methods may be employed to support realization of approaches and mechanisms to meet the challenge of requirements imposed by distributed systems. Usage of process calculi for high-level and abstract system specifications enables early verification of system properties, such as absence of deadlocks or delivery of expected service. Reactive languages provide suitable tools for handling of real-time and synchronization features. Object-based approaches are more inclined to capture the dynamic reconfiguration and structural changes of systems and with proper design may also address the issues of heterogeneity, autonomy, and mobility.

We note that today's distributed applications have a relatively high degree of coupling among the various network resources needed to realize their implementation. This is undesirable because it makes applications much harder to develop and maintain, a situation further worsened when dealing with multiple languages and platforms. As a result, enormous complexity is embedded in the network resources comprising such systems (e.g., communication protocols and object models).

CORBA Distributing Object Computing

A very simple but powerful model for distributed object computing is one that has been adopted by the Object Management Group (OMG) and serves as the foundation for CORBA (Common Object Request Broker Architecture) standards [5]. CORBA deals with distributed object systems by means of a broker paradigm. Objects and clients (which are other objects) might be distributed, but the mechanisms and semantics for object interaction are, from the client perspective, the same regardless of the location of the target object. This is accomplished using an object request broker (ORB), a key element in the CORBA architecture. The ORB is responsible for all of the mechanisms required to find the object implementation satisfying the request, to prepare the object implementation to receive the request, and to communicate the data comprising the request. The interface the client sees is completely independent of where the object is located, in what programming language it is implemented, or any other aspect that is not reflected in the object's interface (Figure 2.1).

We have highlighted CORBA because it incarnates one possible solution for distributed computing that is currently implemented in the marketplace, having wide availability of commercial off-the-shelf products, tools, and components. Further, it has wide acceptance as an industry standard (reflected in its support by more than 900 companies) for distributed computing and systems integration.

2.2.3   Requirements Criteria

As mentioned earlier, the growing demand within the telecommunications industry for improvements in performance and for interoperability of various vendors' equipment and network control and management systems is key. Imagine the challenge of designing a set of specifications that can be used by multiple vendors, across different development teams, to support a customer application. Further imagine doing so in a way that enables sufficient flexibility to support the range of architectures, management protocols, and associated availability and performance desired by each operator in a maximally cost-effective manner.

In the past, the assumption of centralized computing architectures enabled requirements to be imposed on what was largely a given physical architecture that initially included a single stand-alone computing platform to embody all

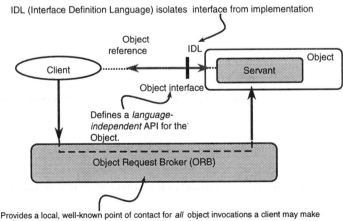

Figure 2.1 A simplified view of the CORBA architecture.

of the software to satisfy a set of requirements. It was clear that the design of the internal software architecture of the computing platform was to be left to the implementers as part of the solution to the problem stated in the requirements and that the requirements should be written in a way to avoid overconstraining the developers. Because early software solutions were generally simple enough to be accommodated, the notion of a software architecture that transcended the system level in order to accommodate the interplay of the functions residing on separate computing platforms was poorly formed since the supposition was that there was one of these per solution anyway. During the first attempts to deal with software solutions involving the cooperation of a multitude of computing platforms, significant effort was expended in standards to arrive at a single physical architecture that would predetermine the physical partitioning of the solution software among the computing platforms. As might be expected, this "one size fits all" solution was rejected by network operators; that is, any rigid partitioning of behavior was considered an unacceptable constraint that would inhibit the ability of network providers to optimally satisfy their network requirements.

We can see the trend for modern software systems to be increasingly open in terms of topology, platform, and evolution, and so the need for a component or object-oriented approach is even more acute than in the past. Even from a single-vendor perspective, the design of these large software systems encompasses aspects that range from the specification of the system's gross structure to the programming of individual components. The system's gross structure, represented as a high-level organization of computational units and interactions between them, is a critical aspect that must be addressed independently of the other aspects. It is important to be able to separate the system properties from the component properties so that system statements can be made on the basis of system structure without involving component information (e.g., avoid deploying system requirements inside components because the resulting components cannot be reused in any other system). To avoid this, it is necessary to utilize system description languages that make it difficult or impossible to mix component internals with system aspects. Additionally, we should be able to check that the overall description of the system is consistent in the sense that the components fit together appropriately. This implies the need for a theoretical basis for describing software architectures that allows us to reason about the behavior of the system as a whole.

From the discussions of the preceding sections, we can make some observations.

- Support for a multivendor environment implies the need for specification mechanisms that provide unambiguous definition of interfaces, without having to specify how functionality is partitioned in particular vendor offerings, or that constrain network architecture choices.

- Because the distributed applications we are considering run on a collection of networked equipment (computing platforms with and without transport interfaces and transport functionality), the application must create the illusion in the minds of its users (including other applications) that it is a single point of computation. Therefore, some of the characteristics that change when the application is running on a single node or on a network of nodes must be transparent to its users. These considerations have significant implications on software and hardware specification approaches. For example, we need to be able to specify the collective behavior of the software components within a group of computing platforms totally independently of the possibilities for internal or external distribution of that software.

- It will be necessary to support distribution and object-oriented design to provide the modularity needed to support a multivendor environment and partitioning of functionality (which may be fixed, provisionable, or dynamic).

- Similar to the object-oriented approach for software specifications, hardware specifications also require a modular approach utilizing a set of basic building blocks, and a set of rules for their association, which should not constrain the equipment design as long as the externally observable behavior complies with the specification.

- Utilization of a variety of technologies, communications mechanisms, and protocols implies the need to foster separation of function from mechanism in specification processes, thus reducing unnecessary dependencies.

- It is necessary to establish requirement traceability that extends all the way from specifications describing the service desired by the customer to the hardware and software specifications associated with the transport and management equipment needed for its realization.

2.3 Conventional Transport Specification Approaches

In general, the specifications associated with phases 1 and 2 described in Section 2.2 have been expressed in terms of natural language constructs (e.g., English and French). The specifications associated with phases 3 and 4 may rely upon signal composition figures to present process connectivity with descriptive text employed to describe the behavior of an equipment for each connectivity case.

Functional analysis and allocation is often performed, where these analyses may be performed via flow block diagrams, timelines, data flow diagrams, state or mode diagrams, behavior diagrams, or other diagramming techniques.

The de facto approach described here has been widely utilized across the telecommunications industry. Within the various network operator and equipment provider organizations, processes and constructs associated with each level of abstraction have been defined. While the generic approach is uniform, it is realized in different ways across the various organizations. Thus, the specific techniques, processes, terminologies, and tools employed by each organization have been different.

Our experience with this approach, and some associated issues, are summarized below.

- This approach appears to make it easier for domain-specific users and specifiers to produce requirements because they do not need to employ any mechanisms that are not within their domain of concern. Each time additional precision is required, text may be added, notes attached to main text, and explanatory figures added, for example. Standards have been established for many years utilizing this approach. However, usage of language-based constructs (pure text-based requirements without formalisms) is well known to offer opportunities for ambiguity and associated confusion and misinterpretation of specifications. It is also known to significantly limit capability for analysis and transformation of a specification into executable code.

- The de facto approach is generally "universally" accepted, though a strict waterfall approach in going from phase 1 to phase 4 is difficult to realize and does not easily accommodate churn. This can become a significant issue when different organizations are assigned responsibility for each phase, with formal "sign-offs" between them. An iterative approach would be preferable, though it would be necessary to introduce some change control as the specifications begin to stabilize for each phase.

- It has been found that the different levels of abstraction can be difficult to enforce because traditionally there have been no uniform mechanisms in place to enforce the levels of abstraction. For example, in an effort to provide unambiguous phase 3 specifications, detail appropriate to phase 4 may be inserted. This can cause issues for phase 4 specifiers because of the inference or assumption of a particular implementation approach. The providers of a phase 2 specification may also unnecessarily constrain the range of architectures and solutions to address the problem domain via their choice of particular features and capabilities.

- Usage of different terminology across organizations (e.g., different language for communicating requirements) can be a significant barrier to reuse of specifications for common functionality. For example, in performing a functional decomposition, different specifiers may group functionality in a different way but use the same term to denote the functional block. Thus, significant time can be spent in just reaching a common understanding of terminology.

2.4 Conventional Management Specification Approaches

In this section, we will introduce and describe the primary management specification approaches in use today. These techniques are the ITU-T Telecommunications Management Network (TMN) specifications, which use the open systems interconnection (OSI) management approach, and the Internet Engineering Task Force (IETF) simple network management protocol (SNMP) approach to management specifications. In this section, we also provide some insight into the degree to which they satisfy their intended objectives and the requirements criteria described previously.

2.4.1 TMN/OSI Approach

In 1979, the International Organization for Standardization (ISO) started work on the OSI framework model, introducing the seven-layer protocol stack. In 1987, the joint group of ISO and CCITT (then ITU-T) began to address standardization of management, which resulted in a set of standards on "OSI Systems Management." During the 1985–1988 time period, CCITT SG 4 attempted to define and develop interoperable multivendor management systems and interfaces for public telecommunications equipment and networks, ultimately resulting in ITU-T Rec. M.3010 [6], "Principles for a TMN." Within the study period extending between 1989 and 1992, ITU-T agreed that OSI systems management would be the basis upon which TMN standards would be built. During the succeeding study periods, since 1992, the ITU-T has continued to enhance TMN and OSI systems management recommendations.

2.4.1.1 OSI Systems Management Specifications

OSI systems management has provided the foundation for most standardization work on telecommunications management. Information on the set of associated standards and their interrelationship is provided in Appendix 2A. The objective of OSI systems management standards was to address the needs of the business community to be able to manage computing and telecommunications networks in a multivendor and multioperator environment. In an effort to achieve this objective, OSI systems management defined the following capabilities:

- A set of standardized conceptual tools for representing (modeling) management information and the resources to be managed;
- A set of standardized means for communicating management information (a standardized management information protocol for the exchange of management information);
- A set of standardized functions and associated means for supporting common management tasks (systems management functions).

The OSI systems management model models the conceptual architecture needed to monitor, control, and coordinate resources within the OSI environment and the protocols for communicating information pertinent to those resources. It generally[3] utilizes an object-oriented architecture where management systems exchange information modeled in terms of managed objects (MOs), which represent conceptual views of resources that are being managed or are there to support certain management functions. We note that managed objects may also represent a relationship between resources or between a combination of resources (e.g., a network). Specifically, network management is considered to be a distributed information application, which involves the exchange of management information between different processes, for the purpose of monitoring and controlling the various physical and logical network resources. For a specific management association, the management processes may take on one of two possible roles:

- *Manager role:* The part of the distributed application that issues management operation directives and receives notifications;
- *Agent role:* The part of the distributed application that manages the associated managed objects. It will respond to directives issued by a manager, and also transmit notifications reflecting the behavior and state of these objects.

All management information exchanges between manager and agent are expressed in terms of a consistent set of management operations (invoked by

---

3. This architecture is object-oriented, except for the functional-oriented concepts of manager and agent.

the manager) and management notifications (filtered and forwarded by the agent). These concepts are illustrated in Figure 2.2.

This exchange of management information is realized through the use of the common management information service (CMIS) and the common management information protocol (CMIP), shown in Figure 2.3.

A "many-to-many" relationship is typical between managers and agents, in the sense that one manager may exchange information with several agents and one agent with several managers. The current architecture for OSI systems management [7] covers only the simple situation where one manager interacts with one agent that in turn communicates with the managed objects (through their manipulation).[4] With respect to the terms *manager* and *agent,* several points are worth stressing.

- Manager and agent are not fixed properties, but are roles taken with respect to a particular management interaction.

- Although the managed objects are shown as located together with the agent, this is not necessarily the case. The requirement is only that the managed objects must somehow be accessible by this agent.

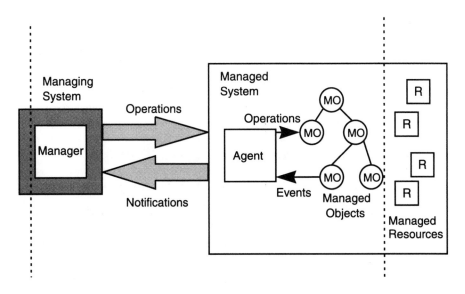

Figure 2.2 OSI systems management model.

---

4. A manager may conduct operations on several agents in parallel, although the options and choice of standardized means for synchronization and commitment, for example, are still being studied.

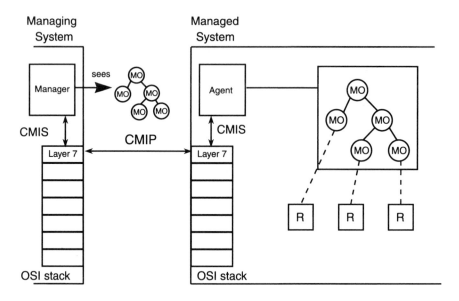

Figure 2.3 CMIP and CMIS.

- The manager and agent must have shared management knowledge. This consists of the collection of managed object classes and managed object instances supported by the agent, the functional capabilities of the agent, and the protocols to use for communication with the agent.

The architecture allows a managed system to take the role as a manager in order to chain operations across systems, as illustrated in Figure 2.4.

The OSI systems management functional areas provide a systematic way of examining requirements to accomplish particular systems management tasks. The five OSI systems management functional areas (FCAPS) are [8]:

- *Fault management* defines the set of facilities that enable detection, isolation, and correction of abnormal operation of managed resources.
- *Configuration management* defines facilities to exercise control over, collect, and distribute data to network resources for the purpose of providing continuous operation of interconnection services.
- *Accounting management* defines the set of facilities that enable charges to be set and allocated for the use of network resources.
- *Performance management* defines the set of facilities required to monitor and evaluate the performance of resources.
- *Security management* defines the set of facilities required to manage the mechanisms that provide access and security protection of resources.

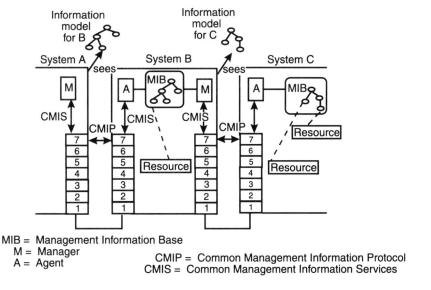

Figure 2.4 OSI systems management architecture.

These functional areas describe broad areas of network management responsibility, and each involves the use of specific supporting systems management functions. Examples of these include object management, attributes for representing relationships, alarm reporting, and objects and attributes for access control. Thus, the specific management functional areas are supported by a set of systems management functions, which in turn are supported by CMIS services, as illustrated in Figure 2.5.

2.4.1.2 TMN Management Specifications

The objective of the TMN standards has been to provide "an organized architecture to achieve the interconnection between various types of operations systems (OS) and/or telecommunications equipment for the exchange of management information using an agreed architecture and standard interfaces" [6]. The functions of the TMN have been categorized consistent with the OSI systems management functions described in the preceding section. A TMN is conceptually a separate infrastructure that interfaces with a telecommunications network at several different points to exchange information and to control network operations. There are four basic aspects or architectures to be considered in the definition of a TMN: functional architecture, physical architecture, information architecture, and logical layered architecture.

The TMN architecture is aimed at specification and development of the interactions between management systems. The specification of TMN, as it is based on OSI systems management, is object-oriented; however, the specifica-

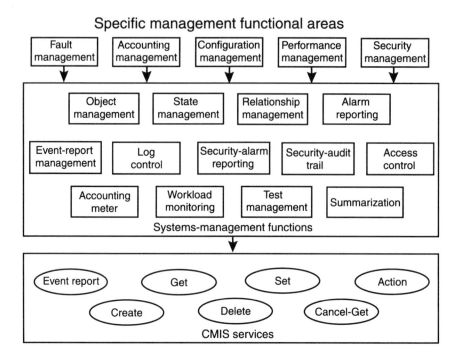

Figure 2.5 OSI management functional areas.

tion of the management information system is functionally oriented at present. The major weakness of the TMN lies in its lack of any significant distribution transparencies and in its large grained specification of physical interfaces. The distributed nature of the system may be hidden from the end users, but all distribution aspects are visible to the applications within the system. Every managed object has a known physical location and it is not possible to move it to a new location. By virtue of the lack of granularity, the interface specification is also inflexible in that new functionality cannot be added as needed. As an example, personal computer (PC) networking applications can be added and removed as desired because each application has an interface specified independently of all other applications. The converse is true for the TMN—all applications using the same physical interface are specified by a single interface.

TMN Functional Architecture

The functionality in a TMN is expressed in terms of TMN function blocks, which include [6]:

- *Operation systems function* (OSF) *block* processes information related to telecommunication management for the purpose of monitoring, coordinating, and controlling the telecommunications network, including the management functions of the TMN itself.

- *Network element function* (NEF) *block* communicates with the TMN for the purpose of being monitored and/or controlled, provides the telecommunications and support functions that are required by the telecommunications network being managed, and includes the telecommunications that are the subject of management.

- *Workstation function* (WSF) *block* provides the means to interpret TMN information for the human user and vice versa.

- *Q adapter function* (QAF) *block* is used to connect, as part of the TMN, those non-TMN (e.g., proprietary) entities that are NEF-like and OSF-like and has the responsibility of translating between a TMN reference point and a non-TMN reference point.

- *Mediation function* (MF) *block* acts on information passing between the OSF and NEF (or QAF) to ensure that the information conforms to the expectations of the function blocks attached to the MF and may store, adapt, filter, threshold, and condense information.

If two function blocks are located in different pieces of equipment, a reference point between them is physically realized by a standardized interface, and denoted by a capitalized letter. This functional architecture and associated TMN reference points are illustrated in Figure 2.6.

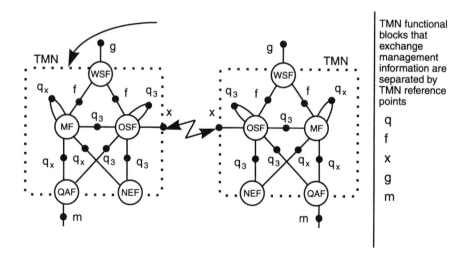

Figure 2.6 TMN functional architecture.

Within function blocks, a number of functional components are defined (such as management application function, directory access function, and security function). The most important of these functional component categories is the management application function, which represents the processing capabilities of the various function blocks (OSF-MAF, MF-MAF, QAF-MAF, and NEF-MAF). Telecommunications management can be viewed by itself as a service, or rather a collection of services. One possible framework for construction of such services, management service decomposition, is given as part of the TMN specifications in ITU-T Rec. M.3200 [9].

TMN Logical Layered Architecture

To deal with the complexity of telecommunications management, TMN management functionality is partitioned into logical layers. The logical layered architecture (LLA) is a concept for the structuring of management functionality that organizes the functions into a grouping called "logical layers" and describes the relationship between them. A logical layer reflects particular aspects of management and implies the clustering of management information supporting that aspect [6]. These layers, illustrated in Figure 2.7, encompass:

- *Business management layer* (BML) has responsibility for the total enterprise.

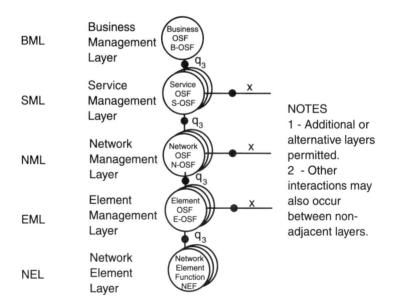

Figure 2.7 Logical layered architecture.

- *Service management layer* (SML) is concerned with, and responsible for, the contractual aspects of services that are being provided to customers or available to potential new customers.

- *Network management layer* (NML) has the responsibility for the management of a network as supported by the element management layer such as control and coordination of the network view of all network elements within its scope or domain.

- *Element management layer* (EML) manages each network element on an individual or group basis and supports an abstraction of the functions provided by the network element layer.

TMN Physical Architecture

TMN functions can be implemented in a variety of physical configurations. The TMN physical blocks represent the physical implementation of TMN function blocks. Each TMN physical block is named by the predominant TMN function it contains. The data communications network (DCN) is a communications network within a TMN that supports the data communications function and represents an implementation of the OSI layers 1 to 3.

TMN standard interfaces correspond to TMN reference points, and as mentioned earlier, interfaces are applied to these reference points when an external physical connection is required. The information to be passed, as identified at the reference point, is captured in an information model for that interface. TMN standard interfaces are defined in terms of the communication protocols (e.g., CMIP, optionally FTAM), the management services (e.g., CMIS and OSI systems management functions), the managed object classes and instances known across the interface, and optionally a QoS and security part. There is a family of protocol suites for each of the TMN interfaces (Q_x, Q_3, F, and X) illustrated in Figure 2.8. The choice of the protocol is dependent on the implementation requirements of the physical configuration. For example, G-type interfaces are for human-computer interactions such as a graphical user interface or a menu where management operations can be selected. F-type interfaces are defined as a collection of basic management function calls. The M-type interface is used to denote non-TMN management interfaces for which a QAF is used. (These interfaces have not all been illustrated.)

Details of the Q_3 family of protocols are given in ITU-T Recs. Q.811 [10] and Q.812 [11]. The protocol suites applicable to the Q_x interfaces may be chosen from any of the ITU-T recommended communication protocols. One candidate protocol suite for the Q_x interface is found in Rec. G.773 [12].

TMN Information Architecture

The TMN information architecture draws heavily on the principles of OSI systems management and represents the formally defined set of network

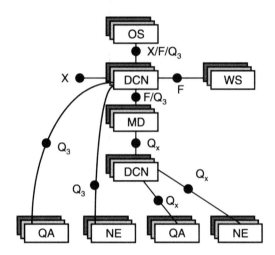

Figure 2.8 TMN physical architecture.

resources and management functions, set of protocols, procedures, message formats, and semantics used for the management communications. The TMN information architecture may be considered from two different perspectives: the management information model and the management information exchange.

The management information model deals with the management aspects of network resources and the related support management activities. It determines the scope of the information that can be exchanged in a standardized manner. Activity around the information model takes place at the application level and involves a variety of information storage, retrieval, and processing functions. Management systems exchange information modeled in terms of managed objects.

Management information exchange involves data communication functions, which allow particular physical components to attach to the telecommunications network at a given interface. It defines communication mechanisms at the TMN interfaces such as protocol stacks. In order to interwork, communicating management systems must share a common view or understanding of at least the following information: supported protocol capabilities, supported management functions, supported managed object classes, available managed object instances, authorized capabilities, and containment relationships between objects.

Traditional TMN information models are based upon the "Guidelines for the Definition of Managed Objects" (GDMO) as described in ITU-T Rec. X.722 [13] and used in M.3100 [14] and its application (e.g., G.774 series Recommendations for SDH [15], I.751 [16] for ATM equipment). These information models describe the information that is made visible (or controllable) by an agent in a managed system to a manager in a management system, as illustrated previously in Figure 2.1. Essentially, these information models serve as the basis for commonality among vendors regarding messages to be sent between network management systems and network elements. Information models can be thought of as a way of organizing data, which can be referenced the same way in messages.[5] For instance, the command needed to change the state of an attribute of an object instance should be the same regardless of the vendor of the equipment. An example of this notation utilizing GDMO is provided in Appendix 2B.

2.4.2 SNMP—Principles and Specification Techniques

In the late 1970s, computer networks had grown from a simple layout of small, separate networks that were not connected to each other to larger networks that were interconnected. These larger networks were called internets and their size grew at an exponential rate. The larger these networks became, the more difficult they became to manage (i.e., monitor and maintain), and it soon became evident that a network management protocol needed to be developed. SNMP was designed in the mid-1980s as an answer to the communication problems between different types of networks. Its initial aim was to be a "bandaid" solution until a better designed and more complete network management approach became available. However, no lightweight better choice became available and SNMP became the protocol of choice for these networks [17].

SNMP was used primarily in customer premises LAN environments to monitor and control data communications devices such as bridges and routers, and between operations centers and the customer premises to provide customers with basic monitoring and control of telecommunications-provided data communications services such as LAN interconnect services. However, its usage has been expanding to encompass management of widely distributed networks.

OSI systems management proposes a clearly stated separation between:

- The management information model (MIM);
- The notation for defining managed objects (GDMO + ASN.1);

---

5. Since information models only describe interfaces, rather than implementations, it is not necessary that either end of the interface actually be implemented in an object-oriented language (even though it might be easier to do so). The only requirement is that the agent system presents what looks like managed objects on the interface and that the manager system interacts with them as if they were objects.

- The definition of management information (DMI);
- The common management information service (CMIS).

SNMP includes much more than simply a communications protocol; while not completely analogous, it is possible to establish an equivalent structure of concerns.

2.4.2.1 Network Management Architecture

SNMP also uses the concepts of managers and agents. Here, the manager is considered to be the console through which the network administrator performs network management functions and agents are the entities that interface to the actual device being managed. These managed "objects" are arranged in what is known as a management information base, also called an MIB, and SNMP allows managers and agents to communicate for the purpose of accessing these managed objects [18]. The model of the network management architecture is illustrated in Figure 2.9 [18].

2.4.2.2 Structure of Management Information

Managed objects must be logically accessible, which means that management information must be stored somewhere and therefore that the information must be retrievable and modifiable. SNMP actually performs the retrieval and modification [18]. This section describes the object model that has been defined

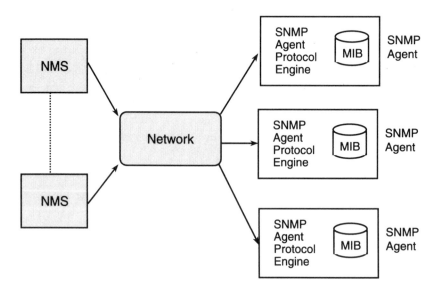

Figure 2.9 SNMP network management architecture model [18].

by the Internet community to represent managed resources. The SNMP structure of management information (SMI) is based on the OSI SMI [18], though it is more aligned with the data model defined by Codd [19] for relational database systems than with the object-oriented model of OSI system management.

The SNMP SMI organizes, names, and describes information so that logical access can occur. The SMI states that each managed object must have a name, syntax, and an encoding. The name, an object identifier, uniquely identifies the object. The syntax (ASN.1) defines the data type, such as an integer or a string of octets. The encoding describes how the information associated with the managed objects is serialized for transmission between machines [18]. Managed object types are described as two-dimensional arrays where each row corresponds to a managed object instance. At most, one managed object instance may be obtained from the instantiation of a given managed object type. Its structure corresponds to a tuple in relational database systems terminology. A tuple is composed of attributes (called scalar objects in SNMP) that all have to be single valued. To simulate multivalued attributes, a single-valued pointer attribute has to be defined that refers to a table, each row of which contains a single-valued attribute. It is worthwhile to note that no action can be attached to object types other than create, read, or write.

Object types may be assembled into groups according to the management information base designer. Generally, grouping is intended for defining a complete set of object types that must all be supported by implementations; grouping may also be made according to a common objective of the object types. In SNMPv2, notification types (defined as traps) can be defined. As in OSI systems management, they are asynchronously sent by the agent to the manager to inform the manager of the occurrence of a specific event related to a managed object. Traps are akin to triggers in relational database systems; they may be defined based on the occurrence of any state change (e.g., attribute value change and tuple creation or deletion). These analogies are provided solely for illustration purposes; an SNMP MIB should not be confused with a relational database, as it represents a virtual object store that might be implemented in a wide variety of ways.

2.4.2.3 Notation for the Definition of Managed Objects

A dedicated notation has been defined to capture the modeling concepts of the SNMP management information model. Syntactical constructs encompass object type, syntax, access (identifies the maximal level of access to the object), status (expresses whether an implementation needs to support the object to claim conformance), and description (helps in reading the object model by providing some explanatory text). An example of this notation utilizing the SNMP-II [20] MIB is provided in Appendix 2C.

2.4.2.4 Definition of Management Information

When defining MIBs, it is often necessary to use similar object types in different places. A number of these object types have been defined and recorded; they correspond to so-called predefined types in some contexts or support objects in other environments.

2.4.2.5 Protocol

SNMP is referred to as "simple" because the agent requires minimal software. Most of the processing power and the data storage resides on the management system, while a complementary subset of those functions resides in the managed system. To achieve its goal of being simple, SNMP includes a limited set of management commands and responses [18]. Contrary to the OSI systems management manager/agent model where manager and agent are roles that can be played by application entities inside any system, the SNMPv2 model considers three types of interactions among them: manager-to-agent, manager-to-manager, and agent-to-manager. For example, the management system issues Get, GetNext, and Set messages to retrieve single or multiple object variables or to establish the value of a single variable. The managed agent sends a Response message to complete the Get, GetNext, or Set. The managed agent sends an event notification (called a trap) to the management system to identify the occurrence of conditions such as a threshold that exceeds a predetermined value [18].

2.4.3 Commentary

In this section, we will examine the degree to which the various approaches we have examined meet the target requirements established earlier.

2.4.3.1 TMN/OSI Management Approach

Apart from the managed objects (from OSI systems management), the TMN/OSI approach is rather functional. The conceptual text of TMN-related standards defines a logical network management architecture and categorizes functions with the intent of supporting the existence of various physical implementations that provide the needed functions. However, they also document protocol recommendations for communications between operations systems and network elements. These protocols are the mechanisms provided by TMN to promote interoperability. TMN communications interfaces are all presently based on GDMO managed object specifications and the CMIP protocol. The protocols are the only enforceable aspect of the standards; all the other aspects are logical or conceptual in nature.

In spite of the object orientation of the specification technique, it is important to recognize that the interface that is actually specified is the complete message set of a single physical interface. While the OSI model makes it possible to specify the minimum necessary at each layer, TMN specifications do not make use of this ability and do not offer low granularity interfaces. In contrast, networked computer systems never specify physical interface message sets. Each application has its own logical interface specification, that is, just the messages relevant to that application. If a new application is placed on the computer there are no changes to the physical interface specification, nor do existing application interfaces change. The new application is orthogonal to the existing applications. TMN specifications ignore the individual applications, yet interoperability can only take place between applications. The equivalent situation in networked computing would be to deny the computer user the possibility of adding applications after the computer had been installed. PC users would clearly view this as intolerable.

At this time, information models have been standardized for SDH/SONET (G.774 series [15]) and ATM equipment (I.751 [16]) and are in the process of being established for optical transport network (OTN) equipment. These models are intended to apply generically to equipment from any manufacturer because the information about the configuration of the equipment is also present in the model. However, conformance to these standards would only indicate that the static definitions of the shared interface are aligned. The bigger problem is to ensure that the total system produces the desired behaviors. This is beyond the scope of such standards, which only provide for informal object behavior descriptions and are confined to a network element view rather than a network view. As a result they provide very little support for the expression of cooperative behaviors. In particular, there is no distribution transparency provided, no support for composition and decomposition of objects, and virtually no support provided for polymorphism (referred to as allomorphism in this context).

While TMN-related protocols promote the concept of multivendor interoperability, they have not, in fact, achieved this result because the protocols are dependent on the existence of a standardized partitioning of functionality between the network equipment and the management systems; in other words, they do not support the flexibility of distribution needed to allow a variety of physical implementations to satisfy the functional requirements. This point is illustrated in Figure 2.10, which illustrates the OSI manager–agent relationship. The agent transfers notifications and operations requests between relatively well-defined managed objects and unspecified management applications. As

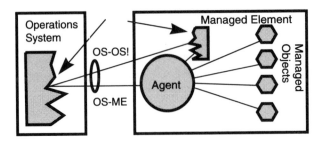

Figure 2.10 A typical split of operations system (OS) functionality.

long as all of the management behavior resides in the remote physical operations system and all of the managed object behaviors reside in the physical network elements, this paradigm basically supports the interoperability requirements (excluding, possibly, the need for the specifications to be unambiguous). However, if for performance, cost, or other reasons, a particular vendor builds some management behaviors into the network element, the manager-agent relationship no longer captures the relationship between the network element and the operations system. Instead, the relationship between the operations system and the network element is actually reflected in the relationship between portions of the management application.

Since the OSI management paradigm does not specify the management application, no standard messages can be derived for communications between the two systems. Further, implementation of a TMN-compliant interface between systems does not assure that one vendor's equipment will behave exactly as another's because the Q3 interface can be used for interactions between management applications, as well as between management applications and managed objects (also illustrated in Figure 2.10). Since these management applications remain unspecified, implementation of a Q3 interface does not guarantee uniformity of behavior. Compounding the problem, templates for the description of managed objects allow only for the informal description of object behaviors. The resulting ambiguities lead to different semantic reactions to messages. Similar problems arise in transport interoperability because of the lack of a common understanding of the applications to be supported by the management overhead built into the transport signals.

The demands of an environment that is flexible in serving the optimization of distribution of network management functionality will eventually exceed the limitations of the rigid partitioning of behavior, with emphasis on the centralization of network management functionality that is implicit in the current TMN standards. This partitioning, as currently interpreted, will inhibit the ability of network providers and manufacturers to provide the technologies needed to respond to these demands. Further, GDMO templates and ASN.1

notation used to specify TMN interfaces are not fully accepted by all elements of the industry. In particular, the formalism has often been judged too heavy for equipment that has limited software capabilities. In the context of global standards, it is clearly advisable to develop a solution that allows a mapping onto any of several well-accepted technologies rather than specifying a single one.

2.4.3.2 SNMP Management Approach

In principle, SNMP is a cheaper version of TMN. It does, however, have one major advantage: the transport protocol used is TCP/IP and it is usual to specify TCP/IP applications at the socket level, which moves the specification to a more logical level than the purely physical interface approach of TMN. Unfortunately, little advantage is taken of this by its "single" MIB design because the single MIB represents all aspects of the managed element. Thus, the fine granularity of specification so essential to interoperability is once again lost.

The fundamental principle in the design of SNMP has been to keep the implementation of agents simple. This principle and the resulting trade-offs made in the design have been important factors in the success of SNMP, making it a low-cost and low-risk choice for managing simple devices. On the other hand, the simplicity of SNMP has its drawbacks.

- Since security algorithms would require significant agent processing, SNMP has only a trivial security mechanism in which passwords are openly passed within each protocol data unit (PDU).

- Since SNMP does not support structured MIB variables, and because implementations need not handle SNMP messages larger than 484 octets, there are no efficient methods for retrieving large volumes of data. Multiple Get-Next exchanges are required, one for each row of a table, which can amount to thousands of messages for large data structures.

- SNMP has technical limitations that make it highly inefficient for higher level, application-to-application exchanges, such as the exchange of provisioning or trouble administration information, where structured queries to large management information bases must be supported.

- SNMP does not support MIB queries based on the content of MIB records. For example, in the SNMP trouble administration MIB defined by a local exchange carrier (LEC), trouble tickets are stored as rows in an SNMP table. However, there is no way to retrieve just those tickets that match some pattern, such as "those still open." The only possibility is to retrieve the whole table and do the selection via processing at the management station.

- SNMP variable naming is assumed to be relative to the agent receiving the request. There is no mechanism for a manager to ask another manager about a variable in a subordinate agent system.

- In typical configurations, SNMP runs over the UDP/IP protocol suite, which is a simple, but potentially unreliable, connectionless transport protocol. This reduces the protocol overhead for the agent but requires the manager to implement time-out and retry mechanisms at the application level. Furthermore, since SNMP traps are initiated by the agent and are unconfirmed, managers cannot depend on being notified.

- SNMP does not scale up efficiently to manage networks with large numbers of network elements because it is used as a polling protocol and must generate substantial network traffic to monitor large networks.

- Several TCP/IP-centric features are visible to applications such as MIB variable names being relative to IP network addresses.

SNMPv2 overcomes almost all of these limitations and provides protocol independence from TCP/IP. However, SNMPv2 does not include an efficient mechanism for selective retrievals from large MIB tables; therefore, it is still unsuitable for applications such as the trouble administration application described earlier. SNMP-based models also cause management system upgrading for each new product introduced into the network because its configuration must be understood beforehand by the management system. Enhancement of SNMP continues with SNMPv3.

2.5 Summary

Network operators are continuing to request enhanced services from the resources of their transport networks, where the enhancements span the range of adding new capabilities (e.g., a wavelength on demand or a virtual meeting room) to improving service performance and availability. They would also like the ability to mix and match vendors' equipment, reduce service introduction time and cost, and reduce the number of interfaces that need to be supported. As we have seen, these drivers put more stringent requirements upon the requirements specifications themselves because they must support distributed applications utilizing a variety of technologies, communications mechanisms, and protocols as well as support unambiguous requirements "hand-off" between different phases of the specification process involving various vendors.

Clearly, two major forces driving improvements in how we express requirements and design specifications are interoperability and the growing recognition that variations in how functionality is partitioned among cooperating platforms can have profound effects on matters such as cost and performance. "All or nothing" physical interface specifications have not been sufficient for the computer industry. How can we imagine that they will be sufficient for the evolving telecommunications industry? Thus, for example, specifications and requirements capture mechanisms will be needed that permit the partitioning of solution software among computing platforms to be included in the architecture of the solution in a manner previously only permitted to the internal software of given physical computing platforms. Further, the notions of precision and independence of problem and solution specification domains take on a much greater importance than they have in the past.

In order to achieve the flexibility that modern networks clearly offer and modern customers clearly want, we have to evolve our specification techniques to focus on fine granularity of the "what" and leave most of the "how" to cost-effective choices made by operators and vendors. In the following chapters, we will explore how this may be accomplished.

References

[1] Lee, R. C., and W. M. Tepfenhart, "UML and C++, A Practical Guide to Object-Oriented Development," Englewood Cliffs, NJ: Prentice-Hall, 1997.

[2] ISO/IEC 10746-21 ITU-T Rec. X.902 "Information Technology, Open Distributed Processing, Reference Model: Foundations," 1995.

[3] Tanenbaum, A. S., "Distributed Operating Systems," Englewood Cliffs, NJ: Prentice-Hall, 1995.

[4] ITU-T Rec. X.901, "Information Technology, Open Distributed Processing, Reference Model: Overview," 1997.

[5] The Object Management Group, "Common Object Request Broker: Architecture and Specification," revision 2.2, Feb. 1998.

[6] ITU-T Rec. M.3010, "Principles for a Telecommunications Management Network," 1996.

[7] ITU-T Rec. X.701, "Information Technology, Open Systems Interconnection, Systems Management Overview," 1992.

[8] ITU-T Rec. X.700, "Management framework for Open Systems Interconnection (OSI) for CCITT Applications," 1992.

[9] ITU-T Recommendation M.3200, "TMN Management Services: Overview," 1992.

[10] ITU-T Rec. Q.811, "Lower Layer Protocol Profiles for the Q3 and X Interfaces," 1997.

[11] ITU-T Rec. Q.812, "Upper Layer Profiles for the Q3 and X Interfaces," 1997.

[12] ITU-T Rec. G.773 "Protocol Suites for Q-interfaces for Management of Transmission Systems," 1993.

[13] ITU-T Rec. X.722 | ISO/IEC 10165-4:1992, "Information Technology, Open Systems Interconnection, Structure of Management Information, Part 4: Guidelines for the Definition of Managed Objects."

[14] ITU-T Rec. M.3100, "Generic Network Information Model," 1995.

[15] ITU-T Rec. G.774 "Synchronous Digital Hierarchy (SDH) Management Information," 1992.

[16] ITU-T Rec. I.751 "Asynchronous Transfer Mode Management of the Network Element View," 1996.

[17] Vallillee, T. (tkvallil@undergrad.math.uwaterloo.ca), "SNMP & CMIP—An Introduction to Network Management," http://www.inforamp.net/~kjvallil/t/snmp.html.

[18] Cohen, Y. (yoramco@radmail.rad.co.il), "SNMP—Simple Network Management Protocol," http://www.rad.com/networks/1995/snmp/snmp.htm.

[19] Codd, E. F. "A Relational Model for Large Shared Data Banks," *Communications of the ACM,* Vol. 13, No. 6, Oct. 1970.

[20] IETF RFC1213, "Management Information Base for Network Management of TCP/IP-based Internets: MIB-II," 1991

Appendix 2A

The set of OSI management standards (ITU and ISO) and their relationship to one another and managed object definitions is illustrated in Figure 2A.1.

Appendix 2B
Example: GDMO Definition of Managed Objects

This example shows the definition of the terminationPoint managed object, as defined in M.3100 [14]. This style of specification is based on the notion of templates, which are actually verbose specification syntax. The type of template (MANAGED OBJECT CLASS and ATTRIBUTE, for example) is spelled out, and in this example the template is a MANAGED OBJECT CLASS specification. Inheritance is provided by the DERIVED FROM clause. Reuse of specifications is provided by the notion of PACKAGES, whereby previously specified items can be imported or items specified here can be exported to other objects. Note that some packages are conditional, which allows some measure of minor customization of the object class. Note, in particular, that the condition predicates in these conditional package statements are plain text and allow no possibility to be computed or otherwise verified. While the idea is good, this failing makes GDMO conditional packages an

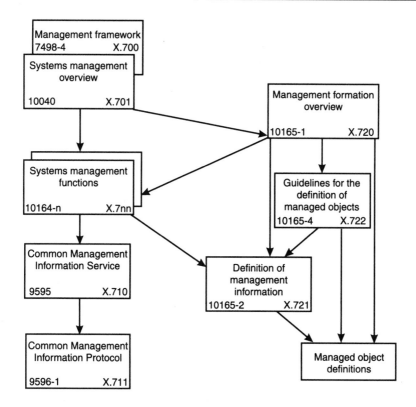

Figure 2A.1 Relationship between ITU-T and ISO standards.

implementer's bane. Similarly, the BEHAVIOUR[6] package, which is intended to specify object class behavior, suffers from the same shortcoming: it is plain text with no possibility of easy verification. The class is closed with a formal Object Identifier, as captured in the REGISTERED AS clause.

```
terminationPoint MANAGED OBJECT CLASS
  DERIVED FROM
    "Recommendation X.721: 1992":top ;
  CHARACTERIZED BY
    terminationPointPackage PACKAGE
      BEHAVIOUR
        terminationPointBehaviour BEHAVIOUR
          DEFINED AS
            "This managed object represents the termination of
            a transport entity, such as a trail or a connection.
            The characteristic information attribute is used to
```

---

6. British spelling is generally used in ITU-T Recommendations.

identify equivalence between subclasses of termination
points in order to determine whether cross connection
or connectivity is possible. The operational state
reflects the perceived ability to generate and/or receive
a valid signal. Subclasses of termination point shall
specify the attributes and states for which attribute
value change and state change notifications will be
generated." ;;
ATTRIBUTES
  supportedByObjectList GET ;;;
CONDITIONAL PACKAGES
attributeValueChangeNotificationPackage
PRESENT IF
  "the attributeValueChange notification defined in Recommendation X.721
  is supported by an instance of this managed object class" ,
operationalStatePackage
PRESENT IF
  "the resource represented by this managed object is capable of assessing
  the ability to generate and/or receive a valid signal." ,
crossConnectionPointerPackage
PRESENT IF
  "the termination point can be flexibly assigned, (i.e. cross connected)." ,
characteristicInformationPackage
PRESENT IF
  "the tmnCommunicationsAlarmInformationPackage package is present
AND the managed object supports configuration of alarm severities" ;
--- several other conditional packages have been removed to avoid clutter
REGISTERED AS {m3100ObjectClass 8} ;

Appendix 2C
Example: SNMP Definition of Managed Objects

This example shows the definition of the object type ifNumber. As described
in the description field, for a managed system this variable captures the number
of different interfaces that are supported by the system. It is accessible from
the manager side in the read only access mode. This variable must be supported
by any implementation that claims conformance to that MIB. Finally, this
object type is uniquely recorded in the sub-tree named interfaces.

-- the Interfaces group

-- Implementation of the Interfaces group is mandatory for
-- all systems.

ifNumber OBJECT-TYPE

SYNTAX INTEGER
ACCESS read-only
STATUS mandatory
DESCRIPTION
      "The number of network interfaces (regardless of
      their current state) present on this system."
::= { interfaces 1 } 2

3

Model-Based Description of Transport Network Functionality

3.1 High-Level Description and General Principles

As discussed in earlier chapters, establishing linkages among services, networks, equipment, and their management is key to realizing the telecommunications vision. To see how these aspects interact, let us consider the "desert start" introduction of a new service by a service provider. Before constructing the service the desired quality of service must be established and a decision made regarding the type of network needed to deliver this selected level of quality. The choices are between a circuit-switched or packet-switched network, between the means of setting up the service by "management" or signaling actions, and among the various technologies capable of supporting the service. Subsequent decisions associated with specifying the chosen network are all related to various aspects of transport functionality. For example, it is necessary to determine how much traffic is anticipated, where that traffic is to be collected from and delivered to, and ultimately to define the supporting equipment requirements. Clearly, in the event that new services are introduced into existing networks, similar linkages must also be established, except that a new range of options must be further considered to determine the optimal approach. These options include reprovisioning or reconfiguration with incremental equipment upgrades, major equipment replacement, or overlay network. In all cases, the support of telecommunications services involves linking three orthogonal dimensions, as illustrated in Figure 3.1: (1) connection-related characteristics, used to support the establishment of connections within networks and equip-

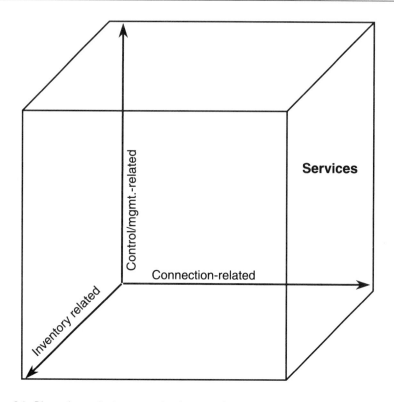

Figure 3.1 Dimensions of telecommunications services support.

ment for transmitting information; (2) control/management-related characteristics, used to control and manage networks and equipment; and (3) inventory-related characteristics associated with "physical realization components," involving knowledge of the underlying hardware and software in supporting equipment.

At the same time, we observe a number of trends and issues for modern networks that significantly impact the network complexity and the planning associated with support of associated services. Transport networking has steadily grown more complex as a consequence of more sophisticated customer needs, the convergence of data and transport networking, and conditions imposed by external market and regulatory forces. In the evolution of embedded core transport infrastructures or in building new core transport networks, efficient cost-effective transport capacity expansion, network reliability, flexible and dynamic bandwidth management, and quality-assured service management are of paramount importance to service providers. Given the wide range of technology choices, there is a trend for networks to employ heterogeneous technology equipment. Whereas in the past transport networks only supported PDH equipment, current networks may utilize varying combinations and proportions of SONET/SDH, IP, and/or ATM equipment as well as WDM/OTN equipment. At the same time, no single physical network architecture seems to exist that satisfies all users' and operators' needs; that is, there are wide ranges of architecture choices. In transport networks, for example, ring, hubbing, and mesh topologies may be needed to satisfy differing applications. A further issue in modern networks has been the lack of a common language that is equally applicable for the description of customer needs, equipment requirements, and standards; that is, there is a need to enable a consistency check amongst business, network, and equipment considerations.

Additionally, current technologies and hierarchies are being designed to facilitate interoperation of equipment produced by different manufacturers, which further widens the competitive aspects of equipment purchase. For example, a supplier of SONET/SDH equipment may therefore supply a number of operators, possibly within a single country, and may be presented with a number of different equipment specifications. At best, this leads to duplication of effort with several, often very comprehensive, specifications relating to the same piece of equipment. In many cases, especially in a period of standards evolution, the specifications each require slightly different functionality, which may reduce competition and increase the price an operator must pay.

From an operator's perspective, the necessity to fully specify a particular type of equipment in order to avoid confusion and misinterpretation by a number of different manufacturers has led to increased specification complexity, which can make it difficult to judge among competing suppliers. Adopting a common methodology for describing such equipment is therefore intended to simplify the specification process, to prevent misunderstanding, and to ensure fair competition. It should also present a set of common basic equipment requirements, facilitating interoperation of multivendor equipment and driving down costs to both the operator and the end user [1].

Taking these factors into consideration, there is motivation for establishing standard model-based approaches that describe and relate the three dimensions (which provide the foundation for telecommunications services), satisfy the criteria described previously, and accomplish the following objectives:

- Enable the description of the generic characteristics of networks using a common language at a level that can transcend technology and physical architecture choices;

- Provide a view of functions or entities that may be distributed among various equipment;

- Concurrently specify transport and management functionality.

Achieving these, in turn, provides critical capabilities to support the following needs.

- Enable the design and planning of networks prior to investments, including selection of the most appropriate types of equipment, to support telecommunications services;
- Facilitate development of transport services and their associated management.

Within standards organizations, work has been done that supports various dimensions of the problem and solution domains. With regard to the dimension addressing transport connection-related characteristics, we can refer to the model-based approach developed within ITU-T Rec. G.805, "The Generic Functional Architecture of Transport Networks" [2]. For the dimension addressing the control and management-related characteristics, we can refer to the model-based approach developed within ETS 300 417, parts 1–7 [3–9]. For the dimension addressing the inventory-related aspects, we can also refer to the "equipment fragment" of ITU-T Rec. M.3100 [10]. The contributions of each part to support the dimensions of the problem will be described in the sections that follow. We will then establish an understanding of the strength of the binding of the dimensions to each other and to the support of the telecommunications services themselves.

3.2 Connection-Related Dimension

A transport network transfers user information from a point to one or more other points bidirectionally or unidirectionally and can also transfer various kinds of network control information such as signaling and operations and maintenance information. Since the transport network is a large, complex network with various components, a network model with well-defined functional entities is essential for its design and management.

ITU-T Rec. G.805 was specifically designed to address the connection-related characteristics of transport networks and equipment. The functional modeling approach it utilizes can be thought of as a requirements capture and analysis tool. Its objective is to describe the information transfer capability of transport networks that is independent of networking technology and to provide a set of "tools" for describing, in a common, consistent manner, the technology-specific transport functionality contained within a complex network. These tools provide:

- A flexible description of transport network and equipment functional architectures;

- A means to identify functional similarities and differences in heterogeneous technology architectures;

- A means to derive equipment functional architectures that are traceable to and reflective of the transport network requirements;

- A rigorous and consistent relationship between these functional architectures and their associated management specifications.

ITU-T Rec. G.805 has been used to provide the methodology and basic concepts that are the foundation of several ITU-T recommendations for technology-specific network architectures:

- Synchronous Digital Hierarchy
    - Rec. G.803, the functional architecture of SDH networks [11];
    - Rec. G.783, the functional architecture of SDH equipment [12];
    - Rec. G.841, the SDH network protection functional architecture [13];
    - Rec. G.842, SDH protection architecture interworking [14].
- Asynchronous Transfer Mode
    - Rec. I.326, the functional architecture of ATM networks [15];
    - Rec. I.732, the functional architecture of ATM equipment [16].
- Optical Transport Networking
    - Rec. G.872, the functional architecture of optical transport networks [17];
    - Draft Rec. G.798, the equipment functional architecture for equipment containing optical transport layers [18].

3.2.1 Basic Concepts

The G.805-based modeling approach has allowed us to analyze the transport network and identify generic functionality that is independent of implementation technology. This has provided a means to describe network functionality in an abstract way in terms of a small number of architectural components, which include topological components, transport entities, transport processing functions, and reference points. These are typically defined by the function they perform in information processing terms or by the relationships they describe between other architectural components. In general, these functions act on a signal presented at one or more inputs and present processed signals at one or more outputs, and are defined and characterized by the information processed between their inputs and outputs. Architectural components may also be associated together in particular ways to form the equipment from which real networks are constructed.

Patterns and structure in the network can be rapidly obscured in a cloud of complex relationships. The connection dimension involves two separate concepts: topology and function. The topology of a network is essentially the set of relationships between nodes (which will later be seen as subnetworks) and defines the available connectivity. Application of this concept simplifies the network description by keeping logical connections distinct from their actual routing in the network and the resources that physically support them. Thus, the logical pattern of interconnection of elements in the network is established without concern for the associated signal processing functions, which allows an operator to easily establish connections as required. On the other hand, the concept of function refers to how signals are transformed during their passage through the network versus how elements of the network are interconnected. Rec. G.805 provides elements that support the modeling of both topological and functional concepts.

Within the topology domain, there are two fundamental concepts that relate to the organization of the network: layering[1] and partitioning.

3.2.1.1 Layering

We have already introduced the concept of topology, which allows us to separate logical connections from the physical routes and resources used to provide them. This logical separation is well represented by the client/server paradigm, where the client refers to the signal being carried and the server refers to the entity providing its carriage; that is, client signals are transported by servers. To utilize this paradigm, we consider the client and server to be two layers, where the client layer is supported by the server layer. The client/server paradigm is recursive, in that any particular server layer could itself be considered a client of another server layer. Further elaborating this paradigm, a network can be represented in terms of a stack of client/server relationships (a stack of layers). It is also useful to note that server layers are relatively more permanent than their clients. This follows from the observation that a server connection must exist both before and after a client connection carried by that server.

Layering therefore enables the decomposition of a transport network into a number of independent transport layer networks, and this independence provides the required separation between its logical topology and physical routes and resources. In particular, the process for setting up connections

---

1. Usage of layering in this context should not be confused with the TMN layers described earlier in Chapter 2.

becomes layer-independent. Layers are also frequently used to define network management domain boundaries within a single service provider's network or between networks operated by different service providers. For example, a layer network can be divided into national subnetworks, which are in turn composed of smaller administrative subnetworks. Network management may also be simplified because each layer's properties can be handled in the same way (e.g., each layer can be assigned a quality of service, monitored for its performance independent of the other layers, and assigned an identification for help in fault isolation).

A layer is defined (characterized) in terms of a set of signal properties, which is called the *characteristic information* (CI) of the layer (e.g., 2.048 Mbps and its format). These are chosen in such a way that any access points having the same characteristic information can be interconnected. This term emphasizes the abstract properties of the stream to avoid the connotations of a physical signal in a medium, though the properties most often chosen tend to be related to the way a particular stream is represented, such as the rate and format at which it transports information. We will return to the more abstract notion of characteristic information later in Chapter 6.

Conventionally, lower order client layer networks use transport services provided by underlying higher order server layer networks. The notion of higher and lower order layers follows the assumption that a server has a higher capacity than its client does; this is sometimes confusing because higher order layers are conventionally drawn at the bottom of the page. The complete set of access points in the layer that can be associated for the purpose of transferring information defines the boundary of a layer network.

For purposes of clarification, we provide an example of layering utilizing PDH (e.g., a DS3 client) and SONET/SDH. The SONET [19] and SDH standards [20] define a hierarchy of signal layer networks (see Figure 3.2) as do the PDH and FDM standards that preceded them. Each layer network requires the services of a higher order layer network to perform the required transport functions. We will discuss the exact location of the layer boundaries in more detail later in this chapter. For this example, we describe the primary signal-layer networks.

- The *logical client signal layer* represents the logical DS3 signal; that is, the DS3 signal rate and format, regardless of physical media characteristics (e.g., line coding).
- The *logical SONET STS* or *SDH VC-3 path layer* network deals with the transport of the DS3 client signal (which may be considered as a "service"). The main function of the path layer network is to provide end-to-end supervision capabilities for the signal, which will traverse

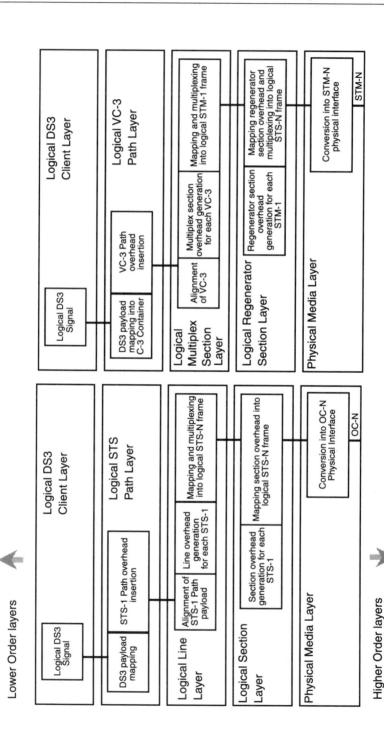

Figure 3.2 SONET/SDH signal hierarchy examples, DS3 client carried on an OC-N/STM-N signal.

a series of SONET line or SDH multiplex sections. Additionally, the layer maps its client into the format required by the SONET line or SDH multiplex section layer network, on whose services it relies.

- The *logical SONET line* or *SDH multiplex section layer* network deals with the reliable transport of path layer network payload and its overhead across the physical medium. The main functions of this layer network are to provide alignment (e.g., frequency or phase) and multiplexing for the path layer network. It relies on the services provided by the SONET section or SDH regenerator section layer network.

- The *logical SONET section* or *SDH regenerator section layer* network deals with the transport of a synchronous transport signal (STS) order N or synchronous transport module (STM) order N frame across the physical medium and uses the services of the physical layer network to form the physical transport. Functions in this layer network include framing, scrambling, and section error monitoring.

- The *physical media layer* network (photonic or electrical) deals with the transport of bits across the physical medium. For example, in the case of photonic media, issues dealt with at this layer network might include optical pulse shape, power levels, and wavelength. This layer is required whenever equipment is to be represented; that is, a physical equipment description is incomplete without provision of physical interfaces.

Thus, using the client/server model recursively, a logical DS3 signal is the client of a logical STS-1 or VC-3 path server. This logical STS-1 or VC-3 path is itself the client of the server logical SONET line or SDH multiplex section server, and so forth.

3.2.1.2 Partitioning

As discussed earlier, the concept of layering helps us manage the complexity created by the presence of different types of characteristic information in current networks, which utilize multiple technologies supporting a wide range of bandwidths. However, even within a single layer, complexity is introduced by the presence of many different network nodes and the connections between them. In order to manage this complexity, we introduce the partitioning concept, which also uses the principle of recursion to tailor the amount of detail needed to be understood at any particular time according to the need of the viewer.

By partitioning, we mean that layer networks can be divided into separate subnetworks that are interconnected by links representing the available transport capacity between them. The role of the subnetwork is to describe flexible connectivity, with no notion of distance being traversed, where traversal of distance is the role of the link. Subnetworks may be delimited according to a wide range of criteria, including those related to network infrastructure, network services, administrative and/or management responsibility, or even geography. Just as a layer network is bounded by access points that can be associated with each other, a subnetwork is bounded by ports that can be associated with each other. (It is important to note that while an access point can only be associated with one layer network, a port may be a member of one or more subnetworks.) Just as layers enable the management of each layer to be similar, so does partitioning allow the management of each partition to be similar.

If we consider that a layer network is actually the largest possible subnetwork bounded by access points, it should not be surprising that subnetworks themselves can also be recursively partitioned into sets of still smaller subnetworks and interconnecting links until the last level of recursion is reached (i.e., a matrix in an equipment). Figure 3.3 illustrates recursive partitioning of a layer network, focusing upon illustrating the principle of partitioning versus the reasons for creating each partition. As each level of partition is created, it is important to understand that the original set of ports around the largest subnetwork neither increases nor decreases in number. The inner subnetworks

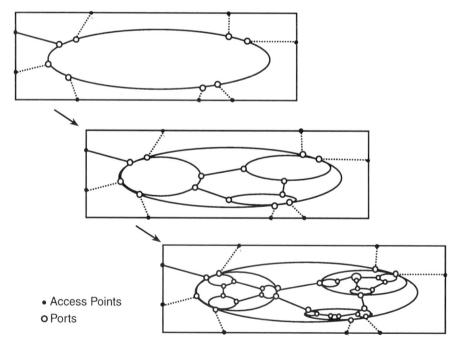

• Access Points
o Ports

Figure 3.3 Recursive partitioning of a layer network.

are intentionally drawn touching the outer subnetworks to indicate that the ports are members of all the touching subnetworks. As more partitions are created, the inner links that become exposed have their own ports on the inner subnetworks. An interesting concept is that at any particular level of partitioning, networks can be considered as a graph, whose vertices are the subnetworks and whose edges are the links.

As might be expected, the rationale for employing a recursive description of subnetworks also applies to links. Recalling that a link can represent available transport capacity between a pair of subnetworks, link connections have been defined within Rec. G.805 as representing the smallest granularity capacity (supported on a server layer) that can be allocated on a link. Thus, a link is considered to be comprised of (partitioned into) a bundle of link connections. However, the concept of link partitioning can be further extended; specifically, we can consider partitioning a link into a set of links of equivalent aggregate capacity (illustrated in Figure 3.4).

This type of link partitioning allows us to assign server capacity to several links rather than to just one. It thus allows us to assign server capacity to several subnetworks, which is necessary for modeling the sharing of a common server layer by several networks. This link partitioning concept is particularly relevant to the modeling of variable-capacity technology networks. From a terminology perspective, links that have been partitioned into a bundle of smaller links in parallel may be considered as compound and component links, respectively (Figure 3.5).

Links may also be serially partitioned into an arrangement of link-subnetwork-link, as illustrated in Figure 3.6; such links may be designated as serially compound and component links, respectively. The usefulness of this concept will be described later in Chapter 6.

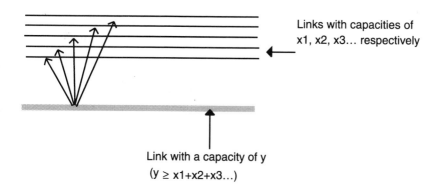

Links with capacities of x1, x2, x3... respectively

Link with a capacity of y
(y ≥ x1+x2+x3...)

Figure 3.4 Partitioning a link into a set of links.

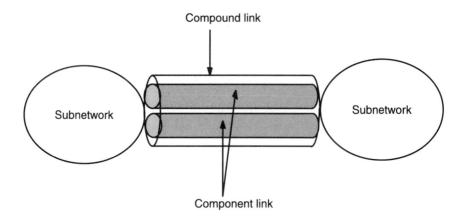

Figure 3.5 Parallel partitioning of a link into links.

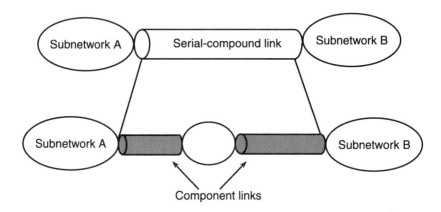

Figure 3.6 Serial partitioning of a link.

The concepts of layering and partitioning are brought together in Figure 3.7, which illustrates a "vertical" arrangement of the layering example described in Section 3.2.1.1. As illustrated, each layer may be considered in terms of a layer network, which can be "horizontally" partitioned into subnetworks to reflect infrastructure or equipment organization, such as self-healing rings, or to reflect convenient management or administrative boundaries. As mentioned earlier, the same network can be partitioned differently for different purposes. For example, the partitioning for connection management of various services, network administration, and maintenance may all be different.

We will close this section with an example of how partitioning enables a network management application to abstract the topology of a layer network (Figure 3.8), which is particularly relevant to the connection management

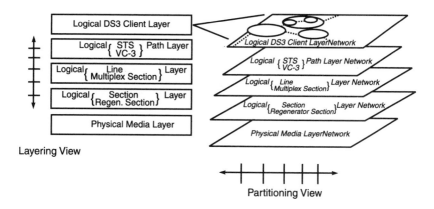

Figure 3.7 Illustration of layering and partitioning.

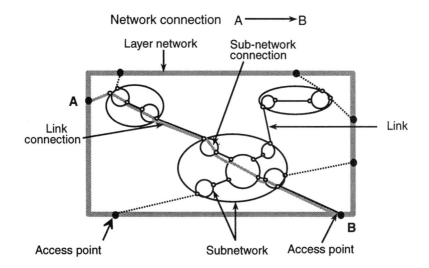

Figure 3.8 Enabling abstraction of the topology of a layer network.

domain. This builds off the example provided in Figure 3.3 and provides a more detailed view of the final stage of partitioning of this particular network to set up a connection from access point A to access point B.

3.2.2 Functionality

While we have discussed the network topology dimension and seen how complexity can be reduced by introducing the concepts of layers and partitions, we have not yet said anything about the functionality needed to actually transport a signal across a network. We have shown how network layers represent client/server relationships; we may consider the functionality involved in transporting signals to be the implementation of these client/server relationships. This functionality is provided by the same small set of transport processing functions in each layer.

The fundamental components of transport processing functionality are known as "atomic" or "elementary" functions and are related to a single layer of the signal hierarchy (or layer network). We will later see that atomic does not mean that the function could not be further decomposed, but rather that we choose not to decompose the function at this particular time (e.g., it is not necessary from the particular layer perspective). There are, however, rules for composition (and decomposition) of atomic functions. Transport processing functions have been identified and grouped into classes corresponding to adaptation and termination. As signal transport is directional, these functions have a source, which originates the signal, and a sink, which receives the signal. Source functions apply a transformation to the signal, and sink functions remove that transformation. Source and sink functions thus occur in pairs within a layer and are bounded by ports, which represent the function inputs and outputs. These ports are actually the same ports that we described as bounding subnetworks and link connections in our partitioning topology model. Transport processing functions are described in more detail in Sections 3.2.2.1–3.2.2.3.

3.2.2.1 Adaptation Function

An adaptation function is an atomic function that passes a collection of information between layer networks by changing the way in which the collection of information is represented into adapted information (AI) that is suitable for the server layer. The adaptation source function is responsible for several key processes:

- *Client encoding,* where the adaptation source adapts a data stream to the server characteristics;

- *Client labeling,* where the adaptation source "labels" each client so that the corresponding adaptation sink can correctly identify it, enabling clients to be multiplexed; however, the means by which this is done is very technology specific;

- *Client alignment,* where adaptation sources align the client signal with capacity in the server layer, while adaptation sinks remove the effects of alignment; although the actual process is technology-dependent, in time division multiplexed (TDM) systems buffering of the signal is commonly required.

3.2.2.2 Trail Termination Function

A trail termination function (TTF) is an atomic function within a layer network where information concerning the integrity and supervision of adapted information may be generated and added or extracted and analyzed. While this function's full title is trail termination function, a common abbreviation is just termination function. The termination source is concerned with creating the ability to monitor signal quality. This frequently involves the addition of components to the signal for the purposes of monitoring, frequently called overhead. The termination sink monitors the signal quality and removes any overhead. It is this overhead removal function that gives the function its name, that is, overhead termination. Overhead can be provided via insertion of additional capacity or, alternatively, usage of already available but unused capacity. The output of the TTF is the characteristic information of the layer.

3.2.2.3 Connection Function

While not a transport processing function, the connection function is a third function in common use. The connection function is an atomic function within a layer, which if connectivity exists, relays a collection of items of information between groups of atomic functions. It does not modify the members of this collection of items of information, although it may terminate any switching protocol information and act upon it. Any connectivity restrictions between inputs and outputs are defined. We note that the connection function is actually the same topological component as the subnetwork and has the same properties.

These atomic functions are represented using a set of symbols, shown in Figure 3.9, that constitutes part of a shorthand diagrammatic notation that will be used for specification purposes. The intent is to simplify technical descriptions via a common set of symbols and naming conventions.

3.2.3 Connections and Points

In Section 3.2.2 we saw that network layer functionality, including the client/server relationship, may be described in terms of a set of elementary functions.

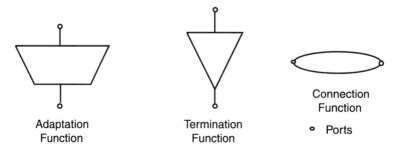

Adaptation Function

Termination Function

Connection Function

o Ports

Figure 3.9 Graphical representation of "atomic" functions.

The client/server relationship itself is most precisely defined as the association between layer networks that is performed by an adaptation function. In fact, these elementary functions can be connected to describe the complete layer behavior, with associated rules describing allowable combinations. Functions are interconnected by considering their ports to be bound together, where a binding between two ports is called a reference point (or just point). This convention makes it possible to illustrate relationships between functions without having to explicitly cite which port is involved. Subnetworks allow flexible bindings between their ports, and the binding of two such ports is called a subnetwork connection. The most commonly used bindings and reference points are described next and illustrated in Figure 3.10.

- Binding of an adaptation source output to a termination source input port is called an access point (AP). This binding is never flexible and can therefore never be partitioned, so it is of relatively little interest. Access points are frequently omitted in functional model diagrams. (An access group is defined as a group of colocated access points together with their associated trail termination functions.)

- Any binding involving a termination source output port or sink input port is called a termination connection point (TCP).

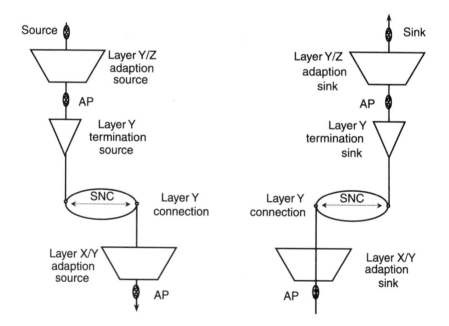

Figure 3.10 Illustration of various bindings and reference points.

- Any binding involving an adaptation source input port or sink output port is called a connection point (CP).

The preceding discussions imply that layers have no "thickness" and are simply planes representing the location of all the connection points in the particular layer. Adaptation and termination functions are located between layers with inputs and outputs in different layers. These "vertical" relationships are usually statically configured while the "horizontal" relationships are usually more dynamic. While this view leads to the least ambiguity in models, conventionally layers are considered to have thickness and the adaptation and termination functions are assigned to either the client or the server layer (Figure 3.11). This convention has more to do with establishing who is responsible for what than with creating good modeling constructs.

3.2.4 Connection Domain Model

We already introduced the concept of a connection as representing an entity that transports information transparently without any integrity control. Several kinds of connections may be identified, depending on the layer and partition traversed by the connection. Some of these were informally introduced, or inferred, earlier in the chapter. They are depicted in Figure 3.12 and more formally defined as follows.

- *Trail:* We saw that access points delimit a layer network. These access points are bound to the input and output ports of trail termination functions. This association between connection termination points is

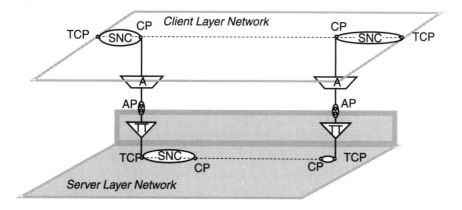

Figure 3.11 Allocation of atomic functions to network layers.

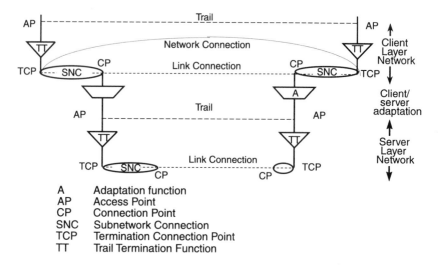

A Adaptation function
AP Access Point
CP Connection Point
SNC Subnetwork Connection
TCP Termination Connection Point
TT Trail Termination Function

Figure 3.12 Illustration of terminology and relationships.

called a trail, and it provides an end-to-end connection offering an automatic means to check the quality of the transport.

- *Network connection:* A network connection represents an association between output and input ports of trail termination functions that transfers information across a layer network without ensuring its integrity. From our earlier discussion of partitioning layer networks, a network connection is comprised of contiguous subnetwork connections and/or link connections.

- *Link:* A link represents the capacity between two subnetworks, two access groups, or one subnetwork and one access group. The granularity of this capacity depends on the implementation technology. Links are both providers of capacity as well as consumers of capacity. A link can be decomposed into several links of lower capacity, each serving different subnetworks or "capacity consumers."

- *Link connection:* A link connection transfers information transparently across a link and is delimited by ports that represent the fixed relation between the ends of the link. These ports are the connection ports associated with an adaptation function.

- *Subnetwork connection:* A subnetwork connection is a transport entity that transfers information across a subnetwork. It is formed by the flexible association of ports on the boundary of the subnetwork. This definition is more specific than that in the G.805 definition, which defines a subnetwork connection as an association between reference points. (The fixed bindings characteristic of trail connection points and connection points may also be thought of as subnetwork connections, which are very often called degenerate subnetwork connections. The subnetwork, or reference point, containing such a connection is very often called a degenerate subnetwork.)

In summary, a trail may convey information for several clients of a layer network by applying multiplexing and transcoding capabilities at the layer network boundary. The existence of a trail in one layer provides any client in that layer with a potential for information transfer between the access points characterizing the extremities of that trail. The client/server relationship, more precisely defined as the association between layer networks that is performed by an adaptation function, allows the link connection in the client layer network to be transported over a trail in the server layer network. The usage of the bandwidth contained in a link is flexible, even if its route may be fixed. Except for the case of definite stable capacity between points that characterize the cable infrastructure, transport services usually involve temporary associations between points. Thus, to allow transport resource reuse, a network needs flexibility (reflected in the subnetwork concept). The potential for flexibility across a subnetwork is governed by an associated policy. The subnetworks give the flexibility and the links give the fixed transport capabilities between subnetworks. Again, as noted earlier, when we refer to fixed infrastructure, we do not mean that it is inflexible; rather, we mean that such possible flexibility is not exercised during the time of the connection we are considering. Links do not change during the time it takes to set up a network connection, neither do the allocated link connections change during the duration of the network connection of which they are a part. In general, the higher the order of the link, the more fixed it tends to be (and vice versa).

The usage of the preceding terminology, and associated relationships, illustrated in Figure 3.12, shows all the relationships (no other arrangements are possible) between ports, reference points, and connections. These restrictions effectively specify a description language.

Figure 3.12 uses the preceding concepts to show how a client layer trail may be transported by a server layer signal. Here, the client trail is first terminated; then transported through a subnetwork, via a subnetwork connection; and adapted for transport across a server layer trail involving server layer subnetwork and link connections. This model allows us to characterize network functionality in a technology-independent manner. In Section 3.2.6 we will provide some examples illustrating application of these principles to specific networking examples involving various technologies.

3.2.5 Sublayers and Function Decomposition

The functions described so far are considered to be atomic at the current level of interest. As with the topological concepts with which we are now familiar, these atomic functions can be decomposed to reveal internal detail when it is necessary. Conversely, more detailed layers can be collapsed to reduce the level of visible detail. The goal, as with the topology models, is to reduce the number of things being dealt with at a given level of interest.

Expanding the adaptation function or termination function (see Figure 3.13) may expand a layer to show more detail. Expansion of the adaptation function allows more detailed specification of the adaptation necessary to create the server layer characteristic information, while expansion of the termination function allows more detailed specification of the termination of the server layer. These techniques have been used to specify greater levels of detail in equipment, new monitoring arrangements in existing layers, fault recovery arrangements for existing layers, and completely new server networks. For completeness, Figure 3.13 also depicts the expansion of the connection point, though this is simply the inclusion of additional resources in the connection.

The converse of expanding layers is, of course, collapsing layers (Figure 3.14). Layers are often collapsed when there are no flexibility points between them and it is not necessary to fully understand the details of every layer. This is most often done in equipment, though it is possible to collapse layers simply to reduce the amount of detail in a drawing.

3.2.6 Examples

Let us first consider how we would model the transport of a PDH DS3 client signal onto an STM-N server signal (Figure 3.15). Here, the logical DS3 client

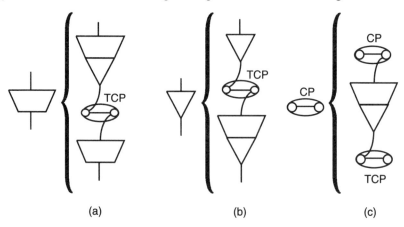

(a)   (b)   (c)

Figure 3.13 Expansion of layers. Expansion of (a) the adaptation function; (b) the termination function; (c) a connection point.

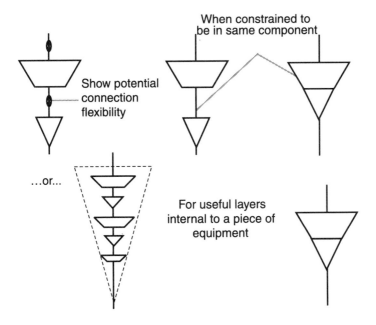

When constrained to
be in same component

Show potential
connection
flexibility

...or...

For useful layers
internal to a piece of
equipment

Figure 3.14 Simplification versus flexibility: collapsing layers.

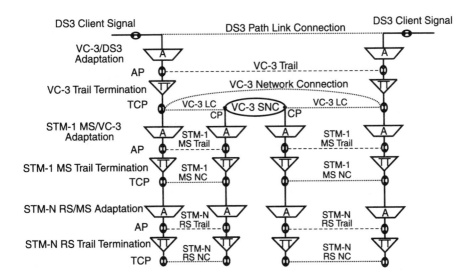

DS3 Client Signal

DS3 Path Link Connection

DS3 Client Signal

VC-3/DS3
Adaptation

A

A

AP

VC-3 Trail

VC-3 Trail Termination

VC-3 Network Connection

TCP

VC-3 LC

VC-3 SNC

VC-3 LC

CP

CP

STM-1 MS/VC-3
Adaptation

A

STM-1
MS Trail

A

A

STM-1
MS Trail

A

AP

STM-1 MS Trail Termination

STM-1
MS NC

STM-1
MS NC

TCP

STM-N RS/MS Adaptation

A

STM-N
RS Trail

A

A

STM-N
RS Trail

A

AP

STM-N RS Trail Termination

STM-N
RS NC

STM-N
RS NC

TCP

Figure 3.15 DS3 client conveyed on SDH VC-3/STM-N server signal.

signal is adapted for transport onto a VC-3 trail via the VC-3/DS3 adaptation function, VC-3 path overhead is provided by the VC-3 trail termination function, and the VC-3 client signal is then adapted for transport on a multiplex section trail (frequency or phase alignment and multiplexing) via the multiplex section adaptation function. Finally, the STM-N regenerator section overhead is provided by the regenerator section termination function.

We also note that it is possible to stop the recursive descent through client/server associations at any arbitrary point. This makes it possible to separate concerns of the different layer networks, enabling focus on the layer network(s) of interest for any particular purpose. For example, Figure 3.16 only describes associations from the DS3 client through the VC-3 trail and network connections, whereas Figure 3.15 shows the remainder of the recursion to the section layers in this example scenario.

The technology and distribution-independent aspects of the functional modeling approach provide a highly flexible tool to accommodate mixed technologies and various possible functional distributions. This is illustrated in Figure 3.17, where we show how a client signal could be carried over an ATM virtual circuit and virtual path, which is in turn carried by an SDH network. (In the figure, the virtual circuit layer is labeled "VC" and the virtual path layer is labeled "VP.") We should note that the virtual circuit and virtual paths are modeled as interlayer functions in exactly the same way as the corresponding relationships in the SDH layers. This gives us a view of the relationship between the client ATM layers and the SDH transport in a completely consistent and integrated manner.

3.2.7 Equipment Packaging

We saw that the topological model of layers and partitions, as well as the interlayer functions, do not specify the packaging of functions into telecommunications equipment. Equipment packaging is the domain where layers, partitions, and functions all come together. We already saw how partitions can be forced by some physical boundary. Equipment provides such a boundary;

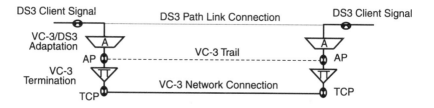

Figure 3.16 DS3 client conveyed on SDH VC-3 server signal.

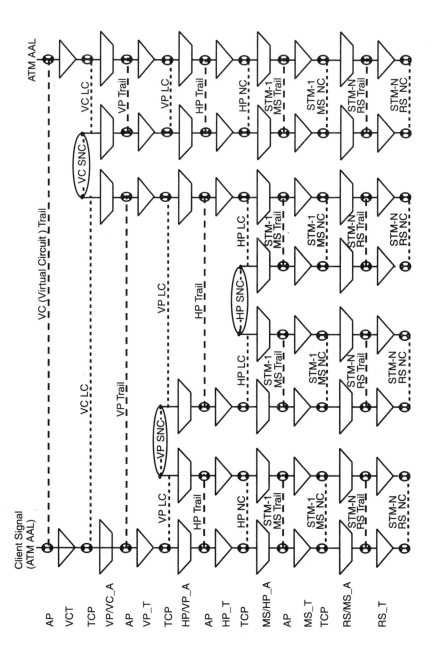

Figure 3.17 Generic architecture involving mixed technologies.

therefore, equipment content is either driven by partitioning decisions or certain partitioning decisions are forced by equipment content decisions. Unlike the network model, which can support logical reference points at any layer, equipment is obviously constrained to only provide physical interfaces. Equipment therefore encapsulates some common element of the layer, partition, and functional models.

It is clear that larger partitions, which are of interest from a network-level perspective, are not usually wholly contained in an equipment. However, as discussed earlier, all partitions are bounded by ports and, since adaptation and termination functions are present only in source/sink pairs, it is clear that any network layer can usually only have one end terminated in any particular equipment. Layer functions also present ports to both client and server layers. Thus, the modeling component that is common from both a network- and equipment-level perspective, is the port. The intersection of the network partition and network layers inside equipment takes place at these ports; that is, equipment encapsulates the ports of a partition and one end of one or more layers. (As a corollary, layers and partitions that are fully contained in equipment are internal matters and are not of interest to a network.) When the equipment allows some flexibility of internal connections, as is generally the case in current equipment, it may be considered to contain an internal flexible subnetwork, which is defined by the ports available for connection (represented as logical resources). Because equipment only provides physical interfaces, all reference points are located inside equipment and are therefore inaccessible. It is this property that allows the functional description of the equipment to be independent of the implementation chosen.

Returning to our example of a DS3 client conveyed on an SDH STM-N signal, we see that Figure 3.15 describes the complete set of functional associations between the client DS3 signal and the logical STM-N signal without ever once referring to any physical equipment. Figure 3.18 shows a possible equipment functional partitioning; that is, a typical organization of functions into equipment—to support transport of the DS3 client signal across an STM-N transport network. Specifically, what is shown is a DS3 connection supported by a VC-3 trail that is terminated by STM-N multiplexers and traverses an intervening cross-connect system with STM-N interfaces and a VC-3 matrix inside. Due to the restriction that equipment can only present physical interfaces, we first complete the model by adding DS3 physical interfaces and ensuring that the STM-N section layers are physical layers.

3.2.8 Application Example

SONET/SDH technology has introduced new network topologies within the transport domain, one of the most notable being the ring. A ring is comprised

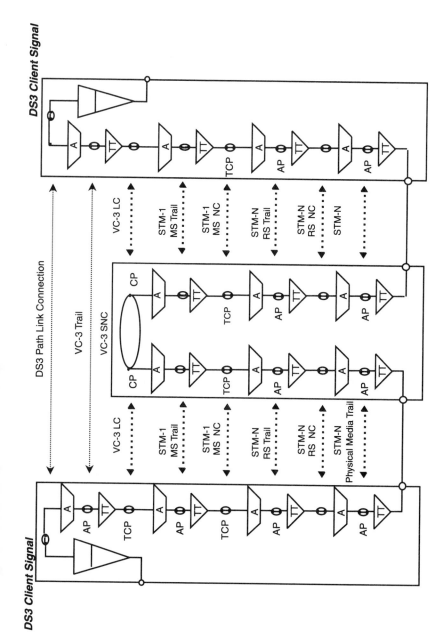

Figure 3.18 Possible equipment functional packaging.

of a set of equipment that allows traffic to enter and to exit the ring, known as add/drop multiplexers (ADM), interconnected in a loop configuration. The main advantage of this topology is survivability: the ring offers traffic two different ways of passing from its ingress to egress (Figure 3.19).

Specifically, the ring's provision of diversely routed fibers allows traffic to utilize an alternative route when its usual route is no longer working. Allocation of pre-assigned capacity between nodes, in conjunction with automatic protection switching facilities, enables the protection of traffic upon detection of a failure. SONET/SDH ring protection architectures significantly improve transport service quality by reducing the unavailable time. Ring protection schemes can provide both trail and subnetwork connection protection. The simplest scheme is 1 + 1 (1 working and 1 protection transport entity) subnetwork connection protection (SNCP) on a physical ring. As illustrated in Figure 3.20 [21], in this case the input traffic is broadcast over two routes (one being the normal working route and the second one being the protection route). As an example, consider the failure-free state for a path from node B to node A. In this case, node B bridges a SONET/SDH path layer signal destined for node A onto both the working and protection routes around the ring (fibers 1 and 2, respectively). At node A, the signals from both of these routes are continuously monitored for path layer defects, and the better quality signal is selected. In the event of a failure, a selector switches to the standby

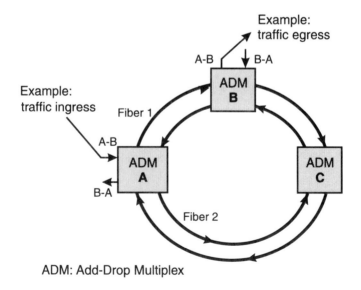

ADM: Add-Drop Multiplex

Figure 3.19 Example ring architecture.

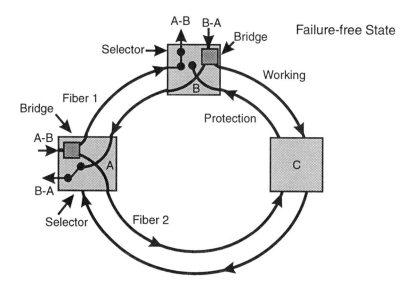

Figure 3.20 1 + 1 SNCP in a physical ring—failure-free state.

at the ring output if the active route fails. This is illustrated in using Figure 3.21 [21] for the case of a failure between nodes B and A [21].

Obviously, the traffic on the ring has a single characteristic information, such as VC-12, and uses the other layers defined for SDH to support these

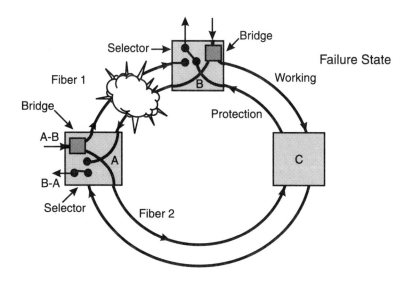

Figure 3.21 1 + 1 SNCP in a physical ring—failure state.

VC-12 connections. We will use this simple example to show how such a network can be modeled from several perspectives; specifically:

- Topological architecture from a network-level perspective;
- Associated transport functions from an equipment-level perspective.

To simplify the example, we will consider the case of VC-12 client traffic being carried by VC-4 server trails. This example could be easily extended by considering the servers of the VC-4 link connections and/or the clients of the VC-12 trails. From the service provider's perspective, the ring may be represented as a VC-12 subnetwork with an associated VC-12 subnetwork connection (illustrated in Figure 3.22).

A given VC-12 subnetwork connection may exit the ring at any equipment; thus, the assignment between the input and the output ports needs to be flexible. This may be supported using the partitioning concept. The previous VC-12 subnetwork may be partitioned into three subnetworks (i.e., equipment fabrics, one in each add/drop multiplex), with the two consecutive ones connected via VC-12 link connections. This level of partitioning is reflected in Figure 3.23.

VC-12 subnetwork

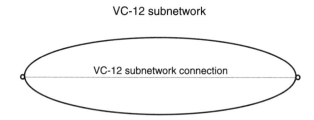

VC-12 subnetwork connection

Figure 3.22 Representation of ring as a VC-12 subnetwork.

VC-12 subnetwork

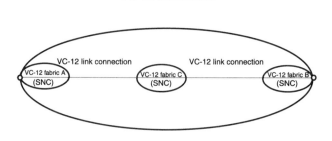

Figure 3.23 Further partitioning of VC-12 subnetwork.

In fact, all VC-12 link connections are served by a VC-4 trail established between the two topologically adjacent equipment fabrics by a process in the server layer, as illustrated in Figure 3.24.

From an equipment perspective, we can now use the network-level model to represent the transport function characteristics that will be implemented in the ADM equipment. If we do so, we see that the equipment characteristics for one ADM may then be expressed as illustrated in Figure 3.25. As mentioned earlier, the equipment models specify a subset of network models that is captured by the physical domain delimited by the equipment implementation part. Only the lowest level of the Rec. G.805 partitioning concept is specified within equipment models.

If we consider the direction from ingress to egress of the ring, illustrated previously in Figure 3.20, we can model the protection scheme in a straightforward manner. The specific application example for a 1 + 1 SNCP ring is illustrated in Figure 3.26.

The selector connection function is flexible and is driven by trail signal fail (TSF) signals derived from the S12_Sk termination points that are reading the S12 layer characteristic information. This arrangement is known as *nonintrusive monitoring* because, while the layer overhead is read to provide signal quality information, the layer is not in fact terminated. We note that these connection functions model the "bridge" and "selector" previously depicted in Figure 3.20.

3.2.9 Application to Packet-Switched Networks

While this section has concentrated upon the circuit-switched aspects of networks (e.g., PDH and SDH/SONET), these concepts can be used to model packet-switched networks (e.g., ATM and IP). In any network, it is necessary to establish an association between endpoints before information can flow between them. The principles for circuit-switched and packet-switched networks differ in one key aspect related to the continuity of the information flow.

- For circuit-switched networks, the route across a subnetwork is decided once for the entire duration of the information flow and the information flow is continuous.
- For packet-switched networks, the information flow is discontinuous (the flow is divided into packets) and the route across a subnetwork is decided on a packet-by-packet and link-by-link basis.

The decision time for the selection of the egress port in the subnetwork is based upon indications that can be pre-established for circuit-switched networks

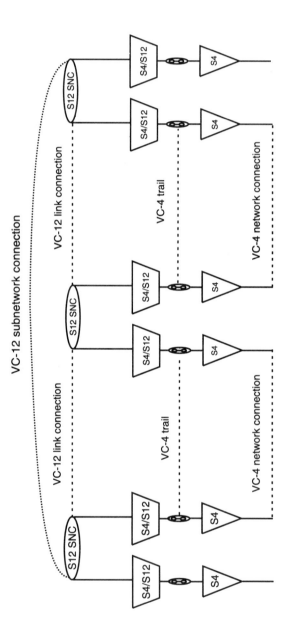

Figure 3.24 Illustration of VC-12 client/VC-4 server relationship.

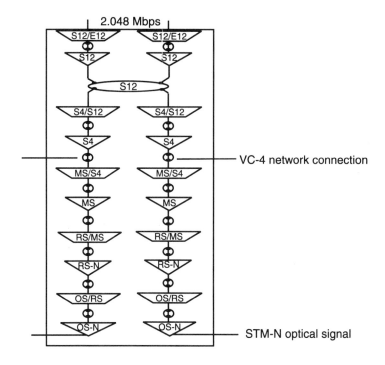

Figure 3.25 ADM equipment characteristics.

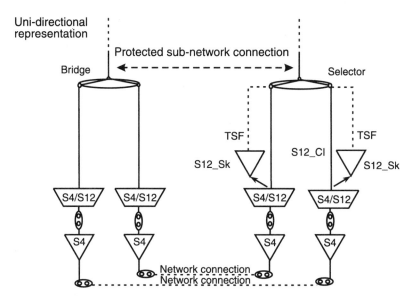

Figure 3.26 Subnetwork connection protection using nonintrusive monitoring.

(existence of a subnetwork connection) or dynamically established for packet-switched networks (based upon an internal subnetwork policy).

The model itself does not make any statements about time scales and presents a cross-section through the network at some point in time. As indicated, in the case of a packet-switched network this snapshot is very short. In order to apply the model to such networks it is only necessary to understand that subnetwork connections can be operated sequentially and that the packet-switched network has a fixed physical topology, just as a circuit-switched network does. Once we can see that the principal difference between circuit- and packet-switched mode networks is one of time scales and sequencing of subnetwork connections, it is clear that all the modeling concepts described in this section are applicable to any network (e.g., as in Rec. I.732 [16]).

3.3 Control and Management Dimension

The functional model of an equipment describes the way in which the equipment accepts, processes, and forwards information contained in a signal. However, functional modeling of the connection-related dimension does not address control and management-related transmission signal processing operations that take place as a client signal passes between layer networks. Functional modeling of the control and management dimension is addressed via the modeling extensions utilized within ETS 300 417 series [3–9], which describes the task of each function from a management perspective; it is an equipment model from a management perspective. Not only are the internal processes that implement the various kinds of equipment functions described but so are their associated internal and external interfaces. Performance criteria that must be met by each process and actions that must be taken when these performance criteria are not met are also specified.

Transmission and equipment supervision processes are concerned with the management of the transmission resources in the network and involve the equipment functionality. The description of these processes requires a functional representation of equipment that is implementation independent. We assume that the manager of the transport equipment has no knowledge of the internal equipment implementation, so equipment faults are recognized as the unavailability of the affected functions. Most atomic functions monitor the signals they are processing for certain characteristics and provide performance information or defect information based on those characteristics. Therefore, transmission supervision processing provides information about the external interface signals that are processed by equipment. Equipment supervision processing is concerned with the fault localization and repair of the equipment itself. Its purpose is to answer the classic question "who to send where to repair what?" with a single replaceable unit at a single location. This desire to report only a single fault stems from the need to avoid overloading management systems and personnel with unnecessary information. It does not require a knowledge of the transmission network, other than that the equipment faults may have been categorized to indicate the severity (e.g., prompt, deferred, maintenance event information) of the fault.

Just as we noted that the network topology can be described recursively (e.g., subnetworks can be partitioned into smaller subnetworks), we can also describe equipment recursively. Specifically, equipment is considered to be built from containers (i.e., racks, which contain shelves, which contain plug-in units) where the limit of equipment recursion is a replaceable unit. As its name suggests, a replaceable unit is the smallest piece of equipment that can be replaced and is usually a synonym for a plug-in unit. It is also the lowest level of granularity needed for equipment fault reporting. Since a manager has no knowledge of the internal details of equipment, it clearly makes no sense to burden the manager with fault reports about details not visible to it. This is an important point to keep firmly in mind when working with the functional model because many functions are frequently packaged on the same replaceable unit and it can be very tempting—but not useful—to consider reporting detailed fault information for each function.

Figure 3.27 illustrates atomic function information points, that is, information exposed by a function that is not part of the payload or its standard overhead. Information points include management, timing, and remote points that generate management information (MI), timing information (TI), and remote information (RI), respectively. Timing and management points can connect to any atomic function. (An independent functional model should be developed for a timing information network. Though we will not describe TI further in this chapter, ETS 300 417-6-1 [8] and G.782 [22] contain timing atomic functions.)

Further detail on function inputs and outputs is illustrated in Figure 3.28. Here the vertical flows represent the payloads, designated as characteristic information (CI) and adapted information (AI).

3.3.1 Equipment Supervisory Process

In this section, we introduce the basic concepts, terminology, and constructs necessary to utilize equipment supervisory process concepts.

3.3.1.1 Basic Concepts

Establishing network- and equipment-level functional models according to the principles described in Section 3.2 provides us the necessary foundation for

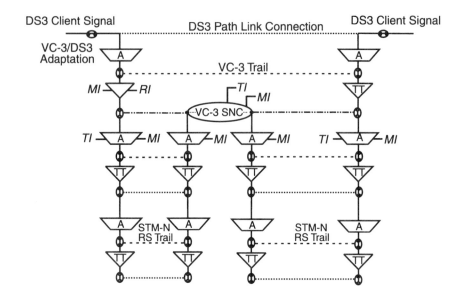

MI - Management Information
TI - Timing Information
RI - Remote Information

Figure 3.27 Atomic function information points.

understanding the concepts associated with equipment supervisory processes. In particular, we will address the processing that takes place within an atomic function to derive basic network management information from the signal and its preparation for passage to the element-level processing that derives alarms and performance parameters. Processing within an atomic function encompasses such steps as detailed definition of the information to be made available to a management system and detailed definition of the parameters that must be provisionable by an operator or management system. Element-level processing refers to the correlation and analysis of information provided by several atomic functions to provide a reduced amount of higher level information to the operator. This is the processing that leads to alarms, performance reports, and lighting of indications on the equipment. In summary, this data is the most detailed available on client signals. The supervision process describes the way in which the actual occurrence of a disturbance or fault is analyzed with the purpose of providing an appropriate indication of performance and/or detected fault condition to maintenance personnel.

In general, performance monitoring involves the continuous collection, analysis, and reporting of performance data associated with a transmission

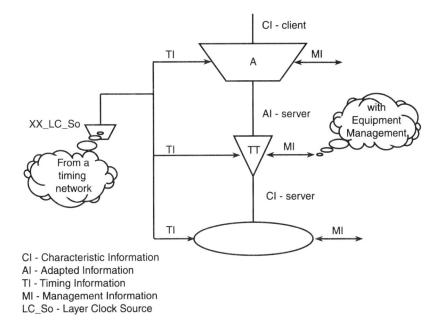

CI - Characteristic Information
AI - Adapted Information
TI - Timing Information
MI - Management Information
LC_So - Layer Clock Source

Figure 3.28 Atomic function inputs and outputs.

entity. It refers to the set of functions and capabilities necessary for equipment to gather, store, threshold, and report performance data associated with its monitored transmission entities. These performance-related data elements are termed performance parameters. Performance monitoring parameters can be generated in trail termination, adaptation, and connection functions. The actual signal parameters monitored are technology-dependent.

Performance parameters are normally gathered under in-service, nonfailure conditions and are typically accumulated (and stored) over predetermined accumulation periods. Performance history data is useful for verifying customer trouble reports and in responding to alerts, so as to quickly assess the recent performance of transport systems (determine QoS) and to sectionalize the trouble or degradation (for example, to locate sources of intermittent errors). This history can also be used in performance assessment against long-term performance objectives [23]. The equipment management function performance monitoring process collects the events associated with trail, link, and protection performance parameters. It counts the events to derive the performance parameters and stores these parameters for later retrieval.

In general, fault (or alarm/status) monitoring is a process that tracks failure events to contribute to an understanding of the overall transmission performance of an entity. The information conveyed via alarm/status monitoring consists of a set of indications that are maintained by the equipment. The equipment sets and clears indications according to well-defined criteria based upon the occurrence and duration of specific events. Some events immediately lead to indications, while others must persist for a specified amount of time prior to the setting of an indication. Alarm and status indications are generally reported under failure events. Alarm/status monitoring and performance monitoring complement one another [9]. Fault monitoring parameters can be generated in trail termination, adaptation, and connection functions.

Each atomic function and performance monitoring process generates and delivers a fault cause to the element management function (EMF) fault management process and a performance indication to the EMF performance monitoring process. The EMF fault management process within equipment performs a persistency check on the fault causes before it declares a *fault cause* a failure. The failure is reported via an output failure report, and by means of alarms (audible and visible indicators). These functions of the supervision process are illustrated in Figure 3.29.

3.3.1.2 Terminology and Constructs

The following terms are used to describe the supervision process: *anomaly, defect, fault, fault cause, failure, consequent action,* and *alarm.* Performance primitives are basic performance-related occurrences detected by monitoring the signal, and these impairment events give rise to various performance parameters. Primitives are grouped into categories of anomalies and defects. Performance parameters are derived from the processing of performance primitives, and the associated terms are defined as follows [9, 23]:

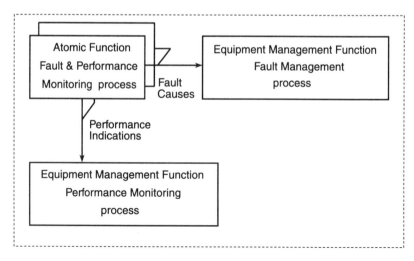

Figure 3.29 Functions of the supervision process..

- *Anomaly:* A performance anomaly is a discrepancy between the actual and desired characteristics of an item. The desired characteristic may be expressed in the form of a specification. An anomaly may or may not affect the ability of an item to perform a required function.

- *Defect:* A performance defect is a limited interruption in the ability of an item to perform a required function. It may or may not lead to maintenance action depending on the results of additional analysis. Successive anomalies causing a decrease in the ability of an item to perform a required function are considered as a defect.

- *Fault:* A fault is the inability of a function to perform a required action. This does not include an inability due to preventive maintenance, lack of external resources, or planned actions.

- *Consequent action:* The action taken in response to an anomaly, defect, or fault.

- *Fault cause:* A single disturbance or fault may lead to the detection of multiple defects. A fault cause is the result of a correlation process that is intended to pinpoint the defect that is representative of the disturbance or fault that is causing the problem.

- *Failure:* Performance failures refer to the termination of an item's ability to perform a required function. At equipment, both local and remote failures can be observed. Local failures involve near-end signal failures, and remote failures are those that occur and are recognized elsewhere and are reported within the transmission system. (While SDH transports most indications in special channels embedded in the signal, this is not necessary for the working of the model.)

- *Alarm:* Alarms are specific types (human observable) of notifications concerning detected failures (or abnormal conditions) usually giving an indication of the severity of the failure. Typically, alarms can be divided into unit level alarms, equipment level alarms, and central office/station alarms.

Detected anomalies are processed in the following manner:

- Anomalies are subjected to a check to identify defects.
- Certain defects initiate consequent actions.
- Defects are correlated to identify the probable fault cause.
- Near-end and far-end defect and performance impairment indicators are counted.

3.3.1.3 Utilization of Concepts

The atomic function fault and performance process can be further expanded to show its internal workings. As an example, the fault and performance processes for the trail termination function are depicted in Figure 3.30. While the details are technology-specific, the basic functions shown are completely generic.

The EMF fault management process is illustrated in Figure 3.31.

Similarly, for the EMF performance monitoring process (Figure 3.32), the individual performance parameter outputs are fed to functions that determine near- and far-end performance according to appropriate standardized performance parameters (e.g., errored seconds) and then forwarded to the performance monitoring history process. As a final step, performance management reports may be provided on a selective basis.

This section has dealt with equipment supervision from the perspective of processes that have been described in equipment standards. In real implementations, the need to reduce the amount of data sent to management systems and personnel leads to internal proprietary mechanisms that are not part of the standard. For example, most switch fabrics use some sort of internal continuity check that is used to generate appropriate fault indications, yet are not part of any standard. In the language of functional modeling, these checks can be modeled as internal trails in exactly the same way as has been discussed for external trails.

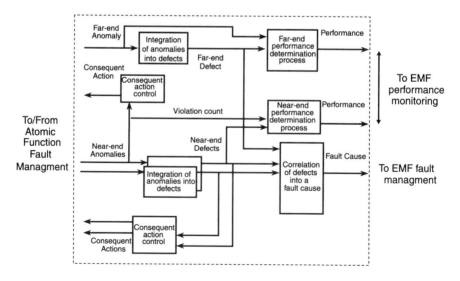

Figure 3.30 Fault and performance processes for TTF.

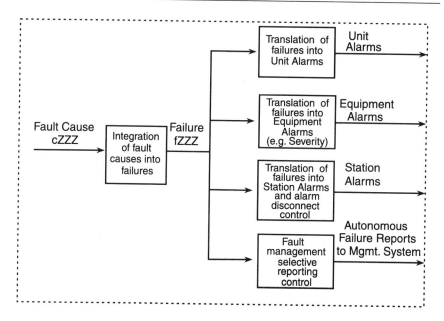

Figure 3.31 EMF fault management process.

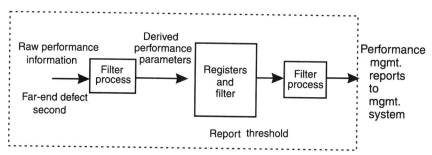

Figure 3.32 EMF performance monitoring process.

3.3.1.4 Application Example

While the atomic function fault and performance monitoring processes are internal to the atomic function, the results of the processing are generally shown as function indication inputs and outputs. An example using the VC-4 trail termination sink function is described next and illustrated in Figure 3.33. In addition to the signal inputs and outputs described in the preceding sections, the termination sink function is considered to have four sets of management related interfaces.

1. An input interface from the server layer that accepts indications about the health of the server layer: these indications can be used to represent a failure or degradation in the server layer network.

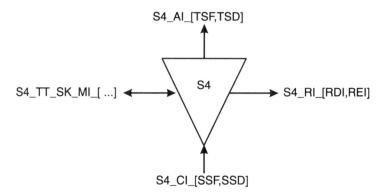

Figure 3.33 VC-4 trail termination sink processing.

2. An output interface to the client layer that provides indications about the health of the trail provided by this layer: these indications can inform the next downstream adaptation function of signal fail or signal degrade condition of the associated trail.

3. An output interface carrying remote indications from the far-end sink termination back to the network element that contains the originating trail terminations source function. These indications convey the status of the characteristic information received by the trail termination sink function, including error detection code violations and other defect information. It is used for single-ended monitoring of the entire trail (i.e., the two extremities).

4. An input/output interface carrying local management indications allowing for the provisioning of threshold data and any local control that may be provided: these indications enable control (start and stop) of the TTF monitoring reporting activities and control of the server signal failure (SSF) report, when detected, to the adaptation function; provide the expected VC-4 trace identifier (the received path identifier has to match with the expected one to prevent misconnections); enable the report of remote defect indicator (RDI) and SSF to the management interface, allowing for the monitoring of the far-end TTF defects from the near-end trail termination function; and allow provisioning of the minimum number of errors, to be detected by this function over a one-second time period, required to declare the second as a bad second.

The naming conventions for these management-related interfaces are [9]:

- Input interface to server layer, <layer name>_CI_[SSF, SD];

- Output interface to client layer, <layer name>_AI_[TSF, TSD];

- Output interface carrying remote indications from the far end sink termination, <layer name>_RI_[RDI,REI];

- Input/output interface carrying local management information, <layer name>_TT_SK_MI_[TPmode, SSF-reported, ExTI, RDI_reported, DEGTHR]; input (or output) signal;

where SSF is the server signal fail, SD is the server signal degrade, TSF the trail signal fail, TSD the trail signal degrade, RDI the remote defect indication, REI the remote error indication, TT_Sk the trail termination sink, TPmode the trail termination performance monitoring control, ExTI the expected trace identifier, and DEGTHR the degraded threshold.

3.3.2 Transport Entities Considered as Management Resources

Network element information describes network elements from a network management perspective in terms of a collection of data that can be manipulated by a management system. Because the management system is more interested in the relationships between layers than in the signal processing aspects of each layer, the functional model is not directly represented to the management system. Instead, the management system looks at an information model that presents data specified in the atomic functions, data computed from this data, and/or data representing relationships between elements of the information model. (It should not be assumed that a one-to-one relationship exists between each function and each data item or that the functional model data is always displayed unchanged.) While the information model is very closely related to the functional model, they are not identical. The functional model is concerned with the transformation of payload signals, while the information model is concerned only with the states of signals for management purposes.

The functional model specifies the most detailed data about the client signals, and the management model describes how this data and derived data is presented to a manager. The management model depends on the functional model; that is, the data to be presented is not fully known until the functional model is complete. We will now introduce a management resource model, and explain how those resources are used from a management perspective, which is described in ITU-T Rec. G.852.2 [24].

The most important resources we will discuss are derived from Rec. G.805 topological and functional concepts. A management resource represents the visibility a manager, or a component for the purposes of management, has in a network. Specifically, Rec. G.805-based management resources allow the manager to "see" what the G.805 topological/functional components allow a network to "do." These resources are enumerated as follows.

3.3.2.1 Connection Termination Point Resource

A connection termination point (CTP) resource encapsulates the management view of the G.805 adaptation function port and the part of this adaptation function that has a 1:1 relationship with the client signal. Thus, it represents the potential end of a link connection and the signal state at idealized points (Figure 3.34).

Because a CTP represents a G.805 port, it can take on the associated topological G.805 relationships. A CTP is always associated with a link connection end and a subnetwork. When the subnetwork represents managed flexibility, the CTP can be bound to a subnetwork connection. When the subnetwork does not offer managed flexibility (the subnetwork is degenerate), the subnetwork connection is not explicitly provided and the CTP is directly bound to the associated CTP, link connection, or trail termination point (TTP). We treat the modeling of degenerate subnetworks in more detail in Section 3.3.2.5.

3.3.2.2 Trail Termination Point Resource

The TTP resource encapsulates the management view of the G.805 termination function port and presents the information derived from the termination function and any adaptation function information that has a 1:1 relationship

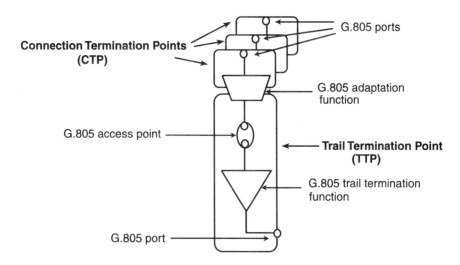

Figure 3.34 Transport component relationship with associated management resources.

with the server layer. It represents an extremity of a trail, is always associated with a subnetwork, and can be bound to other resources depending on whether the subnetwork provides any managed flexibility. When the subnetwork is flexible, the TTP can be bound to the subnetwork connection resource. When the subnetwork is degenerate, the TTP can be bound directly to the associated CTP or link connection resource.

3.3.2.3 Link and Link Connection Resources

The link resource is derived from the G.805 link topological component. A link connection resource is derived from the G.805 link connection topological component and represents the limit of link partitioning, which is the minimum capacity that can be managed. Some technologies can only provide capacity in discrete chunks of large granularity, while others can provide finer grained capacity. Since a link connection resource is a special case of a link resource, some properties of the link resource apply equally to the link connection resource. We note that some network managers only see chunks of capacity between subnetwork resources (link resources) and not the individual link connection resources. For example, in the public switched telephony network (PSTN), connection setup is delegated to an internal connection setup process that uses node-to-node signaling protocols to coordinate the allocation of resources. Thus, the managers of the PSTN network manage topology, while the signaling protocol manages individual connections.

3.3.2.4 Subnetwork Resource

The subnetwork resource is derived from the G.805 subnetwork topological component and is used to effect routing of a specific characteristic information. It follows G.805 topology rules and may be composed of smaller component subnetwork resources that are the result of partitioning. While the G.805 subnetwork is always bounded by a set of ports, the subnetwork resource can be used to associate larger entities. Specifically, the subnetwork resource can associate termination points, link resources, or link connection resources where the entities associated by a particular subnetwork resource depend upon the distribution of management routing functionality. Connections are always made between termination points. However, if the establishment of individual connections is distributed to a different system (e.g., a signaling system for route control, as for the PSTN or in packet or cell-based networks), it is not necessary for a particular management application (e.g., PSTN topology manager) to see each individual termination point in its network view. In this case, that manager would only need to see a subnetwork resource bounded by link resources.

3.3.2.5 Subnetwork Connection Resource

The subnetwork connection resource is derived from the G.805 subnetwork connection, which associates two ports on the edge of the subnetwork. In the resource model, a subnetwork connection resource represents a transport entity that transfers information across a subnetwork resource. The subnetwork connection resource is the component that actually associates the subnetwork resource bounding entities. A subnetwork connection resource can associate edge components (i.e., link connections, links, CTPs, or TTPs) on a partitioned subnetwork resource (Figure 3.35). In this case, the subnetwork connection resource can have an inner structure reflecting the subnetwork resource and link connection resources resulting from that partitioning.

When the subnetwork connection resource associates points on a degenerate subnetwork resource, the subnetwork connection resource itself can be considered degenerate. In the cases where the subnetwork resource is degenerate (allowing no managed flexibility) or is uninteresting to management for a different reason, provision of full management visibility is unnecessary and a modeling burden. Thus, to address such cases, the resource model allows the subnetwork topological component to be replaced by a direct binding between the subnetwork edge components. When the extremities are CTPs, then the CTP pair is collapsed into a single CTP that binds a pair of link connections.

3.3.2.6 Trail Resource

The trail resource is derived from the G.805 trail and represents a transport entity that is responsible for the transfer and integrity of information between two trail termination points. Because a trail associates termination points on the layer network and the layer network can be partitioned, the trail resource

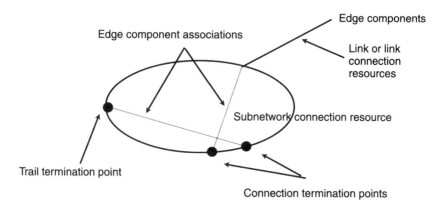

Figure 3.35 Subnetwork connection resource associations.

can have internal structure representing the subnetwork resource and link connection resources resulting from that partitioning (Figure 3.36).

We have seen how management resources are derived from G.805 topological and functional components. In particular, we have management resources for connection points, derived from G.805 ports, and various connections. In the management domain it is possible to describe the network in terms of connection points and the relationships between them or in terms of link connections and the relationships between those. A link connection, after all, is nothing more than a relationship between the connection termination points at the link ends. A model based on link connections is often called an arc view of the network. While there are no hard and fast rules about when the arc view is useful, such a model is often used when the link connection is entirely within a single administrative domain.

3.3.3 Relationship to TMN Equipment Information Models

The TMN model describes network elements from a network management perspective in terms of a collection of managed objects that can be manipulated by a management system. The collection of managed objects, which describes the information that is passed across a physical interface in the TMN, is usually called a management information model. As we will see in Chapter 4, this is perhaps more correctly called a management, or TMN, information model to avoid confusion with other uses of the term *information model*. Chronologically, however, TMN standards used the term *information model* first, so care should be taken to be certain of the context in which the term is used.

These managed objects are derived from the management resource model described earlier. TTPs represent the signal state as it leaves the layer, while CTPs represent the signal state as it enters the layer. These information model points are identical to the CTP and TTP resources described earlier in this chapter. The connection function in the functional model represents either a

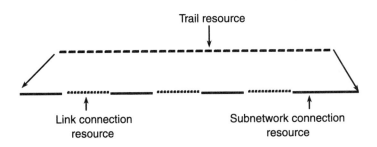

Figure 3.36 Trail resource associations.

fixed or manageable relationship between adaptation functions and is derived from the subnetwork and subnetwork connection resources. In the information model, the fixed relationship is modeled by a fixed relationship between CTP objects and is implemented by *upstream* and *downstream pointer attributes*. A manageable connection relationship is modeled using an additional *cross-connection object* and *fabric object*. The fabric object corresponds to the subnetwork itself (i.e., the matrix) and provides a point to which management actions can be directed. Referring to Figure 3.37, we can provide an example of the relationship between the functional model of a network element and the associated information model.

The information model objects have attributes that represent status and control (supervisory) information that can be associated with the termination and adaptation functions. The TTP object class has attributes that represent supervisory information associated with part of the adaptation functionality and the termination functionality and is assigned an idealized location. Similarly, the connection termination point object class has attributes that represent information associated with most of the adaptation functionality and is assigned an idealized location. The associated termination point subclasses for technology-specific applications can be based upon their functional models. For SDH, these can be derived based upon the SDH functional model and are associated with the SDH path, multiplex section, and regenerator section. Similarly, for OTN, these subclasses can be derived based upon the OTN functional model.

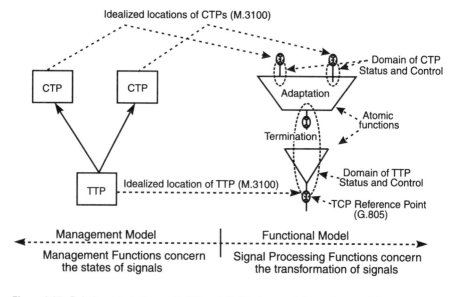

Figure 3.37 Relationship between G.805 and TMN element information model.

3.4 Inventory-Related Dimension

Inventory is considered to be the physical embodiment of the equipment function. It encompasses the software loaded into equipment as well as the more obvious physical racks, shelves, and packs, for example. Since there are no functions in the functional model related to inventory, only information inventory objects exist; these are described in the equipment fragment of ITU-T Rec. M.3100 [10]. As might be expected, a fragment is a more or less self-contained part of a larger complete model. The equipment fragment therefore describes the physical hardware and has little connection with other fragments of the model.

When used to describe a physical implementation, the layer functions described in the functional model certainly do have a relation to the equipment fragment. After all, these functions are implemented by some physical equipment. This relationship is not usually described in the functional model viewpoint and is left to the information objects already mentioned. The main objects of the transport fragment are the termination points, which have a "supported by" relationship to objects of the equipment fragment that allows the relationship between service and equipment to be established. Additionally, processes described in the functional model give rise to equipment alarms, which are only specified in the information model and are generally ignored in the functional model.

The equipment fragment will not be pursued further in this book.

3.5 Summary

Each of the dimensions described earlier has been described via various model-based techniques. In some cases, the linkages are quite strong, such as that between the connection-related dimension and the supervisory processes realm of the control and management dimension. Specifically, the functional modeling method utilized within these dimensions is based on a recursive decomposition technique that decomposes the network into layer networks, layer networks into atomic functions, atomic functions into processes, and processes into detailed requirements. The created set of processes and atomic functions establish a kind of library of components, which can be used to describe many types of equipment. As such, there are no longer any mandatory requirements for particular equipment in standards. A requirement is mandatory for equipment only if the equipment includes an atomic function that contains a process that is stated to support the requirement. It becomes the manufacturers' responsibility to select the proper subset of atomic functions within each of the network elements offered.

However, the linkage between the areas described previously and the management information modeling area of the control and management dimension has remained significantly weaker. In fact, since these areas are traditionally addressed within different standards groups, there has been constant effort to avoid disconnects between functional capabilities specified for networks and equipment and their associated management specifications. Once we begin to examine services management and its associated modeling, we will see that there are further opportunities for significant disconnects between telecommunications services specifications, transport network specifications, and equipment specifications. In fact, there is no direct means for establishing traceability between service-related requirements and specifications of the underlying transport infrastructure and equipment. The challenge is to leverage the strengths of the model-based techniques described in these chapters while establishing a context for strengthening linkages and enabling traceability of requirements from the service level to the equipment level. We will attempt to do this in the chapters that follow.

References

[1]   Brown, S., "A Functional Description of SDH Transmission Equipment," *BT Technology Journal,* Vol. 14, No. 2, Apr. 1996.

[2]   G.805 "Generic Functional Architecture of Transport Networks," 1995.

[3]   ETS 300 417-1-1 "Generic Requirements of Transport Functionality of Equipment, Generic Processes and Performance," Edition 1, Jan. 1996, draft revision, Jan. 1997.

[4]   ETS 300 417-2-1 "Generic Requirements of Transport Functionality of Equipment, SDH and PDH Physical Section Layer Functions," Edition 1, April 1997.

[5]   ETS 300 417-3-1 "Generic Requirements of Transport Functionality of Equipment, STM-N Regenerator and Multiplex Section Layer Functions," Edition 1, June 1997.

[6]   ETS 300 417-4-1 "Generic Requirements of Transport Functionality of Equipment, SDH Path Layer Functions," Edition 1, June 1997.

[7]   ETS 300 417-5-1 "Generic Requirements of Transport Functionality of Equipment, PDH Path Layer Functions," draft for voting, Dec. 1997.

[8]   ETS 300 417-6-1 "Generic Requirements of Transport Functionality of Equipment, Synchronization Distribution Layer Functions," final draft after Public Enquiry, Jan. 1998.

[9]   ETS 300 417-7-1 "Generic Requirements of Transport Functionality of Equipment, Equipment Management and Auxiliary Layer Functions," draft version 0.5, July 10, 1998.

[10] M.3100 "Generic Network Information Model," Equipment Fragment published version, July 1995.

[11] G.803 "Architecture of Transport Networks Based on the Synchronous Digital Hierarchy (SDH)," published version, June 1997.

[12]   G.783 "Characteristics of Synchronous Digital Hierarchy (SDH) Equipment Functional Blocks," published version, April 1997, draft revised version, Feb. 1998.

[13]   G.841 "Types and Characteristics of SDH Network Protection Architectures," published version, July 1995, draft revised version, Feb. 1998.

[14]   G.842 "Interworking of SDH Network Protection Architectures," published version, Apr. 1997.

[15]   I.326 "Functional Architecture of Transport Networks Based On ATM," draft revised version Sept. 1997.

[16]   I.732 "Functional Architecture of ATM Equipment," 1996.

[17]   Draft G.872 "Architecture of Optical Transport Networks," draft version, Oct. 1998.

[18]   Draft G.798 "Equipment Functional Model for Equipments Containing Optical Layers," draft version, Oct. 1998.

[19]   ANSI T1.105-1995, "Synchronous Optical Network (SONET)—Basic Description including Multiplex Structure, Rates, and Formats."

[20]   G.707, "Network node interface for the Synchronous Digital Hierarchy (SDH)," published version, Mar. 1996.

[21]   Manchester, J., and P. Bonenfant, "Fiber Optic Network Survivability: SONET/Optical Protection Layer Interworking," in *Proc. of NFOEC'96,* Denver, CO, 1996.

[22]   G.783: "Characteristics of Synchronous Digital Hierarchy (SDH) Equipment Functional Blocks," Apr. 1997.

[23]   T1.231-1993, "Digital Hierarchy—Layer 1 In-Service Digital Transmission Performance Monitoring."

[24]   G.852.2: "Enterprise Viewpoint Description of the Transport Network Resource Model," Mar. 1999.

4

ODP Architectural Framework Applied to Telecommunications Systems Design

4.1 Selection Rationale

From the discussion of the previous chapters, it is clear that any modeling technique employed must be able to address the range of perspectives (service, network, equipment) that corresponds to the needs of multiple applications for different network operators. Such a modeling technique must also enable separation of concerns and precise specification of each of these concerns, as well as independence from underlying implementation considerations (e.g., communications mechanisms and processing platforms). As will be described in Section 4.2, the reference model for open distributed processing (RM-ODP) [1–4], which constitutes a framework of abstractions for the specification of open distributed systems, satisfies the preceding criteria. Specifically, by enabling a separation of concerns, it provides a means for separating the logical specification of required behaviors (i.e., functional requirements) from the specifications of physical architectures implemented to realize them. This separation of concerns is not intended to imply a lack of real-world constraints or requirements imposed by the physical environment on the functional specification (or vice versa). Rather, it allows the problem domains of physical and functional specifications to be tackled in an iterative fashion on a "divide and conquer" basis rather than forcing parallel and simultaneous concern about the problems introduced by each. This approach also enables functional "needs" to be expressed independent of any particular solution. Thus, it allows us to focus

103

on software applications and their required interfaces without having to immediately consider the various software technologies or notations used for establishing the specification (e.g., ASN.1, IDL, UML, SDL). For example, data networking applications are commonly managed using SNMP, as described in Chapter 2. However, many large public operators prefer to use CMIP, and CORBA is on the horizon. If the application is specified according to RM-ODP concepts, it can be defined independently of any specific protocol but can be mapped to any of them.

4.2 Specification Framework, Architecture, and Modeling Constructs

The reference model for open distributed processing (RM-ODP) defines a number of general modeling concepts (foundations) [2] and perspectives enabling selective focus upon some aspects of the system without needing to immediately consider other aspects of the system (architecture) [3]. On the one hand, it enables focus upon aspects related to externally observable behavior without needing to simultaneously consider infrastructure or architecture-related aspects. On the other hand, it also enables focus upon describing a distributed infrastructure in terms of systems, communication channels, and protocols, for example, without needing to consider applications-level issues. Requirements for a system are specified in a functional manner, independent of any notion of how they might be physically distributed in a particular implementation or what technologies might be employed in developing the system. Distribution issues, as well as technology alternatives that can meet the specified requirements, can then be addressed independently. In the next sections, we illustrate how usage of RM-ODP concepts, augmented with telecommunications domain application methods, enables generation of specifications in a manner that provides for separation of concerns and facilitates their traceability, flexibility, reuse, protocol independence, and conformance testing from every point of view. (We note that RM-ODP in itself should not be considered as an analysis and/or design method and, indeed, has never claimed to be such. In this book, we will not be focusing upon characterizing the differences between "pure" RM-ODP and its application for telecommunications.)

RM-ODP modeling concepts and architecture have been specified in ISO/ITU-T standards [1–4]. The concepts for ODP-compliant open distributed processing systems, sufficient to establish requirements for specification techniques, are defined in Part 2, "Foundations" [2]. To our knowledge, this document is the only existing international standard targeted at finding a solution to the problem of heterogeneity in object orientation. It provides standard definitions for the base concepts that are manipulated by all programming languages and/or analysis and design methods (which often have various object-oriented models). Part 3, "Architecture" [3] specifies the characteristics that qualify a distributed processing system as being "open." This document constitutes an important work in the area of system design, whether the system comprises a single piece of hardware, software running on a single PC, or an entire geographically distributed enterprise with a set of processes and information flows (that may or may not be automated).[1] The application of RM-ODP to telecommunications has been specified in ITU-T standards [5–8]; in particular, the method for using RM-ODP concepts in specifying network management services for purposes of standardization is described within ITU-T Rec. G.851-1[2] [5].

The fundamental concepts of RM-ODP are the viewpoint and the object. A viewpoint on a system is an abstraction that yields a specification of the whole system as related to a particular set of concerns [1]. The RM-ODP architecture enables the design of any system, including systems heterogeneous in their implementation. Within telecommunications, support for heterogeneity in systems design is essential and implies an iterative multistep methodology. The need for iteration arises, for example, from the requirement to express specific implementation constraints that may only become relevant in later stages of the design. Each step in the iterative process may be thought of as a viewpoint.

Objects provide a representation of real world entities and are considered to have existence, lifetime, behavior, and state. It should be noted that the RM-ODP definition of an object is broader than that inferred by traditional object-oriented methods (e.g., restricted to programmatic inheritance). As discussed in Chapter 2, objects may be considered to offer services to their clients. As an informal definition, object state refers to the condition of the object that determines what its current behavior will be (which is strongly related to object attribute values). Unlike mathematical functions, objects have a sense of history. Depending on the particular viewpoint under consideration, more or less emphasis will be placed on object behavior or state. These concepts will be further refined as we describe each viewpoint in detail.

---

1. Part 1, "Overview and Guide to Use" [1] contains a motivational overview of ODP, providing the scope, justification, and explanation of key concepts and an outline of the ODP architecture. Part 4, "Architectural Semantics" [4] contains a formalization of the ODP modeling concepts defined in RM-ODP, Part 2. The formalization is achieved by interpreting each concept in terms of the construct of the different standardized formal description techniques, including Z, SDL, LOTOS, and ESTELLE.

2. G.852-854 specifies particular network management services and associated models according to the method of G.851.

In order to provide different abstractions of any given potentially distributed system, RM-ODP [1] defines five viewpoints (illustrated in Figure 4.1), each of which enables focus upon a particular aspect of the system without needing to consider aspects covered in other viewpoints. While each viewpoint can be treated as disjoint from the others, their union constitutes a complete system specification.

These viewpoints are defined [1] as follows:

- *Enterprise viewpoint* is concerned with the purpose, scope, and policies governing the activities of the specified system within the organization of which it is a part.

- *Information viewpoint* is concerned with the kinds of information handled by the system and constraints on the value changes, use, and interpretation of that information.

- *Computational viewpoint* is concerned with the functional decomposition of the system into a set of objects that interact at interfaces, enabling system distribution.

- *Engineering viewpoint* is concerned with the infrastructure required to support system distribution.

- *Technology viewpoint* is concerned with the choice of technology to support system distribution.

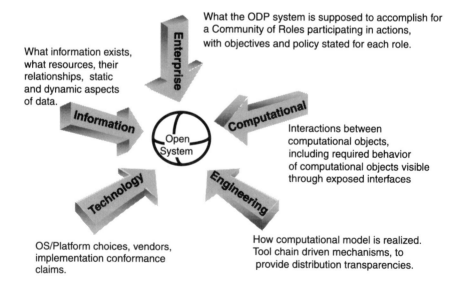

Figure 4.1 RM-ODP viewpoints.

It should be noted that RM-ODP does not force a system designer to use the entire set of viewpoints or standardize any sequence in their use. It rather provides a toolkit made up of these viewpoints where the designer can select the tools (i.e., viewpoints) that are the most appropriate for their needs. Traditionally, top-down specification approaches include a first step whose purpose is to express requirements and/or desired behavior of the system, which is captured by an enterprise specification. Next, information representing the system is defined (information specification), and processes that enable access to and computation on that information are developed (computational specification). These three specifications are made independent of any concerns regarding the ultimate distribution of functionality. The next stage of specification deals with the mapping of these processes onto a possibly distributed architecture. While this specification method is essentially top-down oriented, it can also be applied in a bottom-up manner as, for example, when reverse engineering a legacy system. Figure 4.2 shows how RM-ODP viewpoints are articulated to provide a complete specification of a system when so desired.

A description of each viewpoint specification, associated modeling concepts, and structure is provided in the following subsections.

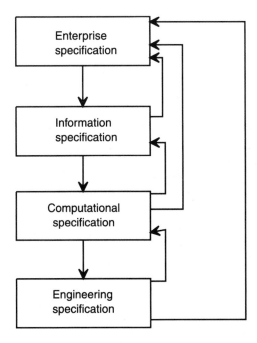

Figure 4.2 Development of viewpoint specifications.

4.2.1 Enterprise Viewpoint

The enterprise viewpoint focuses on placing the system of interest, the enterprise, in its environment. The system of interest, or enterprise, can be macroscopic or microscopic; that is, it could be a large international bank or just one office in a particular operational branch of that bank. The enterprise viewpoint can capture different abstractions of the system of interest. For example, it might describe only what is externally observable about the system, that is, a "black box" description of inputs, outputs, and externally observable behavior. Another abstraction of the same system might consist of describing only its internal aspects, that is, internal components and interactions between them. Both of these abstractions represent valid enterprise viewpoints, though the same information is not captured in each.

4.2.1.1 Scope

An enterprise specification defines the purpose, scope, and policies for the system of interest in terms that are meaningful to the system stakeholders [9]. The system of interest provides a service; thus, it can receive service invocation requests from its environment, deal with those requests, and send back results to its environment. In order to fulfill a requested service, the enterprise may need to work with a number of entities, some of which may be locally resident within the enterprise and some that may reside in another system. Depending on the complexity of the system and the level of detail required for its description, a certain number of entities are involved in providing the service, each one playing one or more roles and providing one or more services. Therefore, the enterprise specification must also describe the relevant entities, their roles, the service(s) they provide, the information flows between them, and their composite behavior.

An enterprise specification can be seen as:

- A document of reference describing user requirements of the service, which can be thought of as the first step of the design of an application that satisfies the client's needs. It provides a template with a fixed structure to clients that allows them to have a clear view of what is required, and also helps in avoiding inadvertent omissions since each concept has to be well defined.

- An official document serving as the basis for negotiations with a contractor providing the service, with the negotiated document serving as a contract between the client and provider. Each clause of the document can be seen as contractual and will constrain the behaviors of both parties.

- A document of reference for testing a supplied system or product, which enables construction of a table listing the different features to be supported. A system claiming conformance to this document would have to support all the features described therein.

- A document representing the requirements of a user that will lead to the definition and specification of corresponding information within the system.

4.2.1.2 Modeling Concepts

An enterprise specification captures the general requirements and/or system usage and management policies in a manner that is completely independent of any implementation or distribution aspects of the system being described. It is up to the specifier to decide whether distribution aspects are to be part of the description of the system (and, if so, to integrate them into the enterprise specification). In addition to the basic modeling concepts of objects and viewpoints described earlier, we will now introduce some specific modeling concepts that are needed to capture enterprise concerns. The metaphor used for the enterprise viewpoint is that of a community of entities that shares a common objective or purpose and provides some service to a client. The entities take on roles in the community where interactions between roles take place according to some set of community policies. Services provided by the community to the client are recorded in contracts.

As an example, let us consider our enterprise to be a bank and discuss its associated community of entities. Clearly, we can consider the employees of the bank to be community entities supporting the service provided by the bank; for example, the service of managing a customer account. However, community entities are not restricted to be active; a bank account itself can be considered part of the community of entities because it is clearly affected by the banking service and affects the outcome of service requests. Each employee of the bank may play a different role in providing the service, or one employee may play several roles at the same or different times. Focusing upon the roles being played rather than on the actual entities playing the roles encourages (though does not require) us to concentrate on what is being done within the community rather than on the entity doing it. In turn, this encourages us to pay more attention to the interfaces between entities than to the internals of an entity. More formal definitions are provided in the following subsections.

Community

A community represents a group of real world entities bound together by an agreement to work together to achieve a particular purpose. As such, a community is characterized by the real-world entities (e.g., people) that play roles in that community and a contract and associated rules (policies) representing the agreements established among them [9]. These entities are referred to as enterprise objects, which are sometimes called business objects in other paradigms. All entities in a community must be relevant to the community purpose (i.e., be relevant to an observer and/or user of the system). It should be noted that entities in a community may also play roles in other communities (i.e., a community is a grouping and there is no abstraction barrier around it).

Roles

An enterprise may be depicted in terms of the roles played by community entities, with each role corresponding to specific responsibilities in providing a service to a client. The service offered to the client is supported by (composed of) interactions between the community entity roles according to the rules of a contract. Two types of entity roles may be distinguished, those of agents and resources. An agent represents an entity able to perform an action, while a resource represents an entity acted upon by an action. Of course, some entities may serve as agents and resources at different times. Among agent roles, there exist two distinct types: the client, which represents the behavior of the caller of a service, and the provider, which represents the behavior of the performer of this service. These two types of agent roles are considered as contracting parties in all service descriptions.

Inside a community, entities may play more than one role; for example, the role of bank teller and supervisor are two distinct roles and may be played by two different persons (two enterprise objects) or by the same person (one enterprise object). The entities playing these roles may also change over time, as when a client is helped by more than one teller. For each distinct role in the community, we need to specify how many instances of that role are either necessary, or allowed to be, in the community. Again referring to our bank example, it is clear that while there is only one teller role, there are many tellers in the bank, though only one may be needed to provide service to a particular customer. While more than one client may be served by one provider, in our specification of a service we always choose a configuration with one client and one provider. The case of several clients or several providers is defined as an extension of the service and is treated separately. This separation allows for a simpler specification of the service by removing the added complication of scheduling these clients and providers.

Contract

The objective of the community is expressed in a contract that defines how it will be realized. Specifically, a service is described in a contract between a user and a provider of the service and results from a negotiation between them.

A specification of a contract contains a description of all the elements of the service, that is, what are the constraints on the provider, what are the constraints on the client, and what is negotiable. In our bank, for example, the user may negotiate the frequency of bank statements when opening the account.

We should note that these contracts could be established at any time in the lifetime of the enterprise. For example, during the lifetime of the bank, policies could change. The bank could decide to send a client a statement every month; offer the client the choice of quarterly, monthly, or weekly statements when the account is opened; or provide a statement whenever the client requests it. (Of course, if the client is unhappy with the terms of the contract, the account could be closed!) In terms of the lifetime of software systems, key choices have to do with whether policy decisions are made at design time, build time, or run time of the system. We cannot stress sufficiently that the choice of which time these decisions are made is a crucial part of the system design.

Some provisions of a contract are necessary, and must be agreed upon, for all clients of a given provider. Others can be negotiated between the provider and each individual client during the establishment of a contract and may or may not be agreed upon. Still other contract provisions can remain optional after its establishment and may be exercised by clients and/or the provider on a negotiated basis or based upon local policy (e.g., such as a "best effort" realization of a service). In order to reflect such policies, specification of an enterprise contract is divided into a two-step process. The first step involves listing all possible contract clauses (in a contract "type") and is used as the basis for the negotiation between a provider and a client to specify, in the second step, which clauses will actually be supported in the contract.

Actions and Activities

While the contract specifies what is to be offered to a client, the community needs a way of specifying the means by which the community satisfies such a contract. Each community has a set of actions that support the community purpose. These actions are used to express the service requests and associated responses that are exchanged between the client and the provider in executing the contract. Again returning to our bank enterprise example, for the "manage account" service we might find actions such as create account, close account, deposit, and withdraw. An action-name and the specification of the action policy define each action.

Frequently, a given behavior is the result of several actions acting in concert and we need a way of specifying these actions and their dependencies. The way this is done is to group these actions into an activity, which is a set of ordered actions (sometimes referred to as a business process). The possible dependencies among the actions are specified by an action graph, which includes a list of atomic actions associated with the activity and a set of activity policies. The action graph and activity policies ensure that dependencies among the actions, such as the ordering of actions within an activity, are clearly specified.

Activities belong to the description of the realization of a service and are very helpful in describing relationships that exist between actions and different services. A particular activity may be specified using standard services but not itself be standardized because that would force a particular solution. It is the responsibility of the systems designer to decide upon the granularity of actions specified. While RM-ODP enterprise actions represent the finest granularity for a given service and are atomic for a given description, they can encapsulate very complex processing. Note that the software system notion of atomic is an operation that succeeds or fails, with no partial results left behind as a result of failure. This requirement for a potentially long sequence of operations to succeed or fail, with no partial results after a failure, substantially complicates the design of complex operations. Such complex processing may be exposed in a more detailed specification, and a corresponding new set of finer grained actions will be defined.

Policy

As in all contracts, contracting parties are governed by the rules of the contract. These rules, constraints, or habits govern the behavior of the whole community, or only part of it, and form what one calls a policy. The policy is specified as a set of applicable rules by either the client or by the provider in the community. It is generally difficult to extract such policies and capture them because some of the policies are so implicitly understood ("it has always been done like that") that they are not expressed. However, an enterprise viewpoint specification needs to capture all policies, including the implicit ones.

Contract rules are defined in concert between the provider and the client and must be accepted by each of them. These rules apply to either one or more of the roles implied in the contract. Three categories of rules have been identified and are called obligations, prohibitions, and permissions.

- An *obligation* rule is a statement that a particular behavior is required and that if the behavior does not occur, the obligation is said to be violated. For example, the bank must remit money to the customer as long as his or her account has a positive balance.

- A *prohibition* is the opposite of an obligation and corresponds to a behavior that must never occur. For example, the bank must not remit money to a customer who does not have an account at that bank.

Should the prohibited behavior in fact occur, then the prohibition is said to have been "violated."

- A *permission* rule is a statement that a particular behavior might occur; it is neither obliged nor prohibited. For example, when withdrawing money from a bank account, a customer may specify the denominations of the currency he or she wishes to receive, while the bank might actually pay with different denominations.

The contract must also describe each case of a contractual violation and its consequence since this is the basis for specifying system error behavior.

4.2.1.3 Structure

The structure of an enterprise viewpoint specification is provided in Appendix 4A. We note that RM-ODP does not propose any specific enterprise viewpoint language or notation but only modeling concepts. It is the responsibility of application-specific domain experts (e.g., network management) to select or design the most appropriate language for capturing their enterprise viewpoint considerations in conformance with RM-ODP concept definitions. Telecommunications experts have defined an enterprise viewpoint language [5], whose notation provides syntactical constructs to capture the concepts described previously in a way that provides adequate rigor of the resulting specification while preserving accessibility of the specification for clients and providers. This notation may be mapped in a straightforward manner to other notations such as *UML use case,* which will be introduced in Chapter 7.

4.2.2 Information Viewpoint

Now that we have introduced a way to describe the enterprise requirements, which are driven by business needs, we need to specify the information needed to fulfill the enterprise concerns. This is done in the information viewpoint.

4.2.2.1 Scope

As its name suggests, an information viewpoint specification is developed entirely on the basis of the information that is available to the system or created by the system, with no notion of computation or of assignment of functionality to any particular processing element. Note that when we talk about information, we mean elements of data; however, it is important to recognize that this is not stored data as in a database system. Neither do we necessarily mean elements of a MIB, the interface specification in the TMN architecture described in Chapter 2. MIB information is only part of the information captured in the information viewpoint specification and is in fact the information that may be manipulated at an interface between two open systems. What is being captured here is the semantics (the meaning) of all the information in the system in a way that is independent of the eventual distribution of that information in any implementation. Another difference from the MIB specification is that the information viewpoint also specifies the way that information is allowed to change, which effectively specifies the behavior of the system. The way that data will be represented (its syntax), stored, and manipulated in an implementation is not relevant to the information viewpoint specification. While the enterprise specification is mainly written by enterprise domain experts, the information viewpoint is usually written by software experts, who specify all the information that is necessary in order to be able to represent the system.

4.2.2.2 Modeling Concepts

The information viewpoint is concerned with the overall system without regard to its distribution. The basic concepts used in its specification are objects, their relationships, and their state transitions. In order to avoid ambiguity, objects used in the information viewpoint are called *information objects.* Transitions between information object states describe the allowed behavior of the system (or subsystem). Because of the focus on state, information objects are considered to be passive data objects; that is, they have type and values but no behavior of their own. Dynamic behavior is captured by the concept of state transitions, where the agent responsible for causing the transition has not yet been specified. This enables a complete specification that is free of distribution and other implementation constraints.

Information Object Classes

The information object class specification encompasses:

- A synopsis of the purpose and behavior of the information object class to aid in understanding the purpose of the object.
- The attributes of the information object class, which can be defined locally or globally.
- The permitted states for an instance of the information object class. These states are expressed as constraints on the value of individual attribute values or a combination of attribute values when the attributes are interdependent. If the specification of permitted values is missing, by default, all possible values for that particular data type are permitted. A state of the object may be valid for just one instant in its evolution or may be valid for the whole lifetime of the object; that is, states may

be used to represent long-term constraints on the object as well as instantaneous constraints.

- The state transitions of the information object define the relevant, valid transitions among all combined states of the information object. These state transitions may be influenced by the existence of invariants (discussed later) that may restrict the visibility of or need for some possible state transitions.

In order to enhance the clarity of the specification, it is not unusual to describe the potential relationships in which an information object can play a role. This set of relationships is not an exhaustive list, nor is it a specification of the relationships that the class, or its subclasses, must support. It is most often used with objects that are not expected to be instantiated but are expected to be subclassed for specific management applications. When such an object is subclassed for a management application-specific information viewpoint, the relationships that are necessary for that application and affect the behavior of the object are listed in the specification of the new subclass and in the relationship parts of the informal description. These relationships may be a subset of those listed for the super-class object as well as any additional required relationships.

Attributes

Just as objects model real-world entities, an attribute models a single characteristic that is applicable to the object type and whose value may vary between object instances. An attribute definition encompasses:

- A synopsis of the purpose and data type of the attribute.
- The allowed states of the attribute, which are essentially all the values that the attribute may take on. When not specified, the allowed values are those of the data type of the attribute. For example, an attribute may be declared as type "Integer" and, with no constraints, the attribute could take on any integer value. The range of values may be explicitly constrained (e.g., 0 to 255) or further constrained by other attribute values, in which case a schema is used for its specification.
- All the possible transitions between the states of the attribute. For example, defining a transition table that constrains the permitted transitions can do this. If this definition is not present, then all state transitions are allowed.
- If an attribute is independent of the others (i.e., its value does not influence the state transition of the other attributes), its transitions

can be described independently. All state transitions that are not described are permitted for this attribute.

We note that in RM-ODP, attribute definitions only provide the semantics of their values, not their syntax. In our telecommunications work, we take as much advantage as possible of the existing TMN work by using existing GDMO managed object models (e.g., G.774) when possible. In this case, the information viewpoint specification imports the semantics of the attribute but not its syntax (which can later be imported into the corresponding computational viewpoint).

Information Relationships

Just as with people, no role is an island and all roles depend upon other entities in order to play some role in the system. These role interdependencies are called relationships and have specific meanings. Because the same relationship is frequently used repetitively in the system, we consider relationships to be classes, and relationship instances are created from the more general relationship class specification.

The relationship class specification is composed of:

- A synopsis of the purpose and behavior of the relationship class;
- The roles that can take part in instances of the relationship class;
- The constraints, called invariants, applicable to the information objects playing the roles of the relationship during the lifetime of this relationship;
- The constraints on the state transitions of the information object instances taking part in the relationship.

Schemas

A schema provides a structured framework for organizing information about states and is used to define more complex system states than only objects or relationships. A schema describes states that involve one or more objects and their relationships, or one or more attributes pertaining to one or more information objects. A system state is expressed by object existence and attribute value constraints within a single object or among several objects. When several objects are involved, the constraint is described in terms of the relationships in which those objects are involved.

The schema definition is composed of:

- A synopsis of the purpose of the schema;
- The roles involved in this schema;
- The constraints that fully define the state.

It is useful to distinguish between the different types of schemas.

Invariant Schemas

The invariant schema is used to define a set of states for a system that is true for the whole lifetime of the system. In general, when we specify an invariant on any collection of objects, we mean that the invariant is true for the whole lifetime of the objects subject to the invariant. It specifies constraints on the states those objects may be in and relationships in which they are allowed to be involved. The invariant schema may involve many objects in the distributed system and specifies conditions to be verified at any time.

Static Schemas

The static schema is used to define a set of states for a system at one particular instant in time. Contrary to the invariant, there is no notion of state before or after this time. This is used to describe a system state before a global system transition (e.g., preconditions of a computational operation, which will become clearer in the discussion of the computational viewpoint in Section 4.2.3.3) and a system state after this transition (e.g., postconditions of the computational operation). Attribute value dependencies and constraints may also be described in static schema and used in the information object specification by referring to the static schema. If the attribute values are further constrained by the values of attributes of other objects, then a static schema must be used. It is only necessary to specify those static schemas that are involved in a dynamic schema describing a valid transition. However, defining any other static schema is not forbidden.

Dynamic Schemas

A dynamic schema is used to express the valid transitions between a static schema defining the state before a transition and a static schema defining the state after the transition. The starting schema is called a precondition, and the ending schema is called a postcondition. It is possible to specify any precondition by referring to an existing static schema, or when the static schema only involves attributes of a single object, the static schema may be specified directly within this dynamic schema. In order to enhance understanding, a dynamic schema specification usually has an informal synopsis of its purpose and behavior. In particular, in the information specification we are not concerned with the cause of state changes, and it is therefore free of unintended implementation constraints.

We can now use the schema concept to rigorously specify system behavior.

4.2.2.3 Structure

The structure of the information viewpoint specification is described in Appendix 4B. Again, just as for the enterprise viewpoint, RM-ODP does not propose any particular notation to capture information concepts and there are several candidate notations including object-based notation [5]. Which candidate is actually used is less important than selecting a candidate that allows us to represent all the necessary semantics (meanings). In Chapter 7, we will illustrate how to use UML as an information viewpoint language and notation.

4.2.3 Computational Viewpoint

The purpose of a computational specification is to define functional units that may potentially be distributed over a number of different computing engines. To achieve this goal, computational object and interface types are defined independently of distribution aspects. These concepts are usually familiar to software application programmers; however, one must be careful to separate the logical RM-ODP computational viewpoint concepts from the implementation concepts of programming languages, formal description techniques, and analysis and design methods.

4.2.3.1 Scope

As we have seen, the information viewpoint is where the system is considered as a whole. Within the computational viewpoint, we begin to define structures suitable for distribution among data processing nodes. (The actual nodes are dealt with in detail in the engineering viewpoint.) These structures, called *computational objects,* are related to the information objects of the information viewpoint, though a single computational object rarely models a single information object. A computational object packages closely related information. We will discuss what makes information "closely related" later in this chapter.

4.2.3.2 Modeling Concepts

The system is seen as a set of computational objects that may be remote from one another and that cooperate with each other. In a system "things that happen" are called *actions*; we are particularly interested in things that happen between computational objects, which are called *interactions.* While purely logical, it is useful to consider an interaction as an unconfirmed unidirectional flow of information from one computational object to another. What is usually more interesting is when this interaction results in a response from the other object. This particular combination of two interactions is called an *interrogation,* a kind of operation, and allows for confirmed two-way communications between objects in the system (Figure 4.3(a)) [3]. We will discuss communications primitives in more detail later in this chapter.

A *computational interface* is defined by the set of operations that may occur, together with the service provided by the interface. This service description is generally known as the behavior of the interface and specifies any constraints, including temporal constraints, on the operations defined by the computational interface. It is important that we avoid confusing an RM-ODP computational interface, which is a virtual interface, with a TMN or OSI interface, which models a physical entity. A TMN interface involves engineering decisions, such as the communications protocol and data representation "on the wire," which are built into the interface. On the other hand, an ODP computational interface specifies only the semantics of the data to be transferred. As the aforementioned engineering decisions are specified later, the interface can be considered as an abstraction (virtual interface). Computational objects cooperate with each other, interacting via their computational interfaces (Figure 4.3(b)).

The behavior of a computational object is defined by its interfaces and the constraints among them. Thus, a computational object provides a means of packaging closely related interfaces into a unit of distribution or component. In addition to its interfaces, a computational object may also provide a specific view of information defined in the information viewpoint. Since the computa-

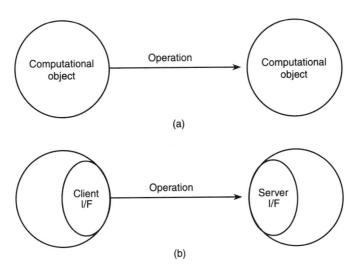

Figure 4.3 (a) An operation between two computational objects and (b) operations between computational interfaces.

tional viewpoint defines only one possible organization of the system into distributable components, there are many possible computational specifications, each describing the same system and each providing for different component distributions. We note that it is easy to confuse the potential for distribution, which is limited by the computational specification, with a particular distribution of functionality that is only specified in the engineering viewpoint. Distribution is limited by the decision to provide (or not provide) a computational interface to a particular information object. Of course, even when such an interface is provided in the computational design, it is still possible to further limit distribution by specifying a nondistributable interface in the engineering viewpoint.

We have suggested that computational objects are related to information objects and that this is not a one-to-one relationship. For example, it would be very unwise to split the control of a complex information object over several computational objects because of the previously alluded to difficulty of providing atomic operations on that data. For the same reason, it would be unwise to split the control of information objects affected by the same operation over several computational objects. However, it would be very usual to provide a separate computational interface to each element of the information object as a way of enhancing the information hiding aspects of the computational design.

Different abstractions of a system can be made at the computational viewpoint. The highest level of abstraction consists of defining only computational interface types. A lower level of abstraction consists of describing the actual implementation of a service by defining the computational object types that support the interface types and interactions and expressing how required behaviors are assured by the implementation. The ITU-T transport network management service standards only specify computational interface types, so choices of component distribution are left to implementers and are not dictated by the standards.

In a computational specification it is not practical to define every instance of each computational object in the system. Thus, the specification defines computational object types, which serve as templates (or patterns) for actual computational objects. Creating an actual computational object is known as *instantiating* an object from its type. As we have already seen in the object-oriented paradigm, we organize objects into an inheritance hierarchy from the most general to the most specialized. The more specialized computational objects (commonly called "children") inherit the properties of their inheritance parent. Just as templates for actual object instances describe computational object types, templates for actual interface instances also describe computational interface types. This allows us to use all the principles of the object-oriented paradigm to organize computational interfaces. In particular, computational interface types are organized as a hierarchy whereby a more specialized interface type inherits the properties of its more generic parent (i.e., super-type). These properties also include the behavior of the interface type so that the interface subtype (commonly called "child") can be considered to behave as its parent interface type. The child interface type may thus be used in relationships that specify the parent interface type, so the more general specification can be met by the more specialized instance.[3] Just as with any object, this property is called polymorphism, which will be very important when we discuss operational parameters later.

In order for actual communication to occur between computational interfaces, a connection must be established between them. Establishing this physical communication is called binding and may be explicit, allowing for control of the communication, or implicit. The time of binding is critical; *explicit* binding occurs at run time, while *implicit* binding occurs before run time and cannot be altered once the system is running.

We have seen that the computational viewpoint describes the system in terms of interacting computational objects and that these objects interact across interfaces. It is necessary to understand the terminology more fully in order to completely describe computational interactions, operations, and interface types. As we have seen, an interaction is the unconfirmed flow of information from one component to another between a pair of bound interfaces. While this is sufficient to describe the communication between components, more detail is required in order to understand what is involved in binding two interfaces. From the point of view of the component, communication takes place between peer components. Within a component, that communication is implemented by giving the information to be communicated to some internal entity that will look after the data transfer (we will later see that this entity is actually the adaptation function of Chapter 3). This internal action is considered to be communication between the sending object and its communications infrastructure, followed by a later communication between the same infrastructure and the receiving object.

Communications between the computational object and its engineering infrastructure are called *signals*. An object with a signal interface acts either as

---

3. While not precisely synonymous, the terms *type* and *class* are frequently used interchangeably in common usage. When referring to the specification process, we will adopt the term *type* to describe the common properties of all the objects pertaining to the same class. In the implementation world of programming, specification types are frequently called classes. In the programming world, a type is frequently considered to be the characteristic of a data element (e.g., integer and array of integers—data type). An object class is considered to be the code that creates the object behavior.

the initiator of the signal submitted to the engineering infrastructure or as the responder to the signal received from the engineering infrastructure. This level of detail is necessary, for example, in order to relate the computational viewpoint to the engineering viewpoint and to discuss error recovery scenarios involving lost messages in the engineering infrastructure. An operation can now be modeled in terms of signals, which enables explanation and modeling of multi-party bindings, end-to-end quality of service characteristics, and bindings between different kinds of interfaces (Figure 4.4).

As we have seen, the operation interface allows communication between two objects. The transfer of a request from the client object to the server object is called the invocation interaction, while the transfer of a result from the server to a client is called the termination interaction. Operations that have output parameters have both invocation and termination interactions and are called *interrogations,* while operations having no output parameters only use the invocation interaction and are called *announcement* operations. Figure 4.4 illustrates the signals and interactions involved in an interrogation operation.

While objects can support both client and server interfaces, all operations at an interface must have the same role (client or server). An interface cannot be a client at one time and a server at another time. Figure 4.5 shows three interacting objects, where one object supports both a client and a server interface. Of course, this does not constrain interface types, which can be instantiated in either the client or server roles on the same or different objects.

So far we have viewed an interaction as being the one-time transfer of a single piece of information. In the real world there are information flows that are continuous. Once started, the information flow continues until it is stopped. Obvious examples of information flows are the transfer of a large file or encoded speech. RM-ODP offers a way to model such continuous transfer

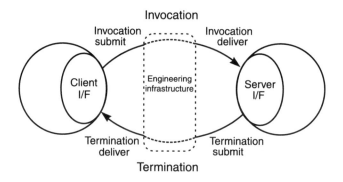

Figure 4.4 Operations between computational objects (viewed as two interactions and four signals).

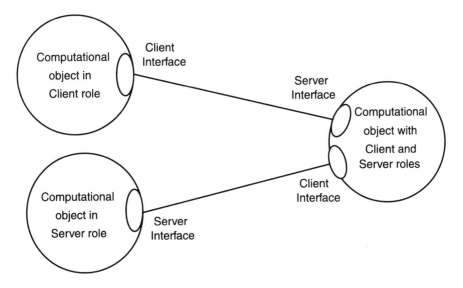

Figure 4.5 Computational objects with client and server roles.

by means of a sequence of interactions, appropriately naming this abstraction as a flow, where interfaces supporting the flow abstraction are called *stream interfaces*. Once started, a flow continues until it is paused or stopped by its producer object. The details of stream interfaces are modeled using signals. Figure 4.6 shows a flow between a producer stream interface and a consumer stream interface. While not especially useful for the majority of management systems, flows provide us with a connection to the payload signal actually being transported through telecommunications equipment. As we will see in Chapter 6, the stream interface is a critical element in modeling transport signals and will be important when we relate functional and management models.

Operations are defined by their input and output parameters, possible exceptions, and pre- and postconditions. The set of input and output parameters and exceptions is known as the *operational signature*. While not actually required by RM-ODP, in our application to telecommunications, we restrict operations

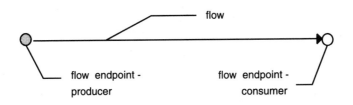

Figure 4.6 Flow interaction between producer and consumer endpoints.

to only reference information that has been defined in the information viewpoint; in this way the viewpoints are more consistently bound together.

- The precondition of an operation is the system state immediately before the operation is invoked,[4] while the postcondition is the system state immediately after the operation completes. This is the point where information viewpoint concepts, such as invariants, and information specifications, such as schemas and data types, are bound to the computational viewpoint. Pre- and postconditions of an operation can be expressed directly in terms of invariants or by references to static schema already defined in the information specification.

- An exception will be raised that causes the termination of the operation if a pre- or postcondition is not satisfied in the execution of an operation. Since there are no standard errors defined, all exceptions must be explicitly specified. To raise an exception on a parameter-matching failure, an invariant referencing the parameter-matching rule is needed.

- Operation parameters, in combination with parameter-matching rules, will select which object(s) will be addressed by the operation in the precondition and which object(s) will be provided as the result of the operation in the postcondition. As might be expected, operation parameters are constrained to be of the type specified in the operational signature.

Parameters can be passed in two ways, by value or by reference, and it is important to understand the difference between them. Stated simply, a parameter that is passed by value does not allow the server to affect the value of the client's parameter in the client environment and represents the "safe" way of doing things. Passing a reference to the parameter does allow its value in the client's environment to be affected by the server. When the parameter is a large or complex structure, it can be more efficient to pass the parameter by reference, though great care must be taken if it is not intended that the server affect the actual value in the client's environment. As an example, passing by value is paying for an item with cash. An amount of money is handed over and there is no possibility of later alteration of the payer's account. Passing by reference is similar to paying by check, in which case the payer's account will be changed at a later time, hopefully in accordance with the amount on the check.

---

4. The usual software engineering meaning of precondition is that which the operation can assume about the client behavior and has no need to check itself. In this case we mean a constraint to be computed by the server itself.

The object-oriented paradigm forces all data to be viewed via an interface, so it is not possible to specify references to information objects. Parameters that refer to information objects can therefore only be value parameters. It is, however, possible to pass complete computational interfaces as parameters, leaving it up to the operation to access any data that is visible through that interface and instructing the operation to obey any access rules for the parameter (illustrated in Figure 4.7).

To allow for complete generality, an interface reference can also be used as a parameter in the operation signature. The type of information object passed must match the type specified for the parameter in the operation signature specification. When an interface reference is passed, the actual interface type must also match the parameter type. In this case a subtype is considered to match, as the more specific interface inherits the properties of the specified interface, and interface polymorphism allows it to behave as the more generic interface type that was specified (illustrated in Figure 4.8). In this case, the

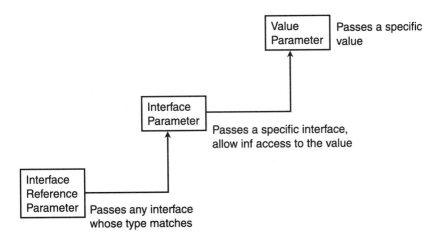

Figure 4.7 Passing computational interface reference parameters.

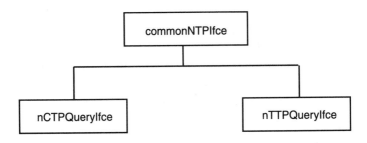

Figure 4.8 Example interface hierarchy.

type constraints may be checked at run time, so the actual interface signature must contain sufficient information to enable this check to be performed.

The run-time value for a parameter is further constrained by a *parameter-matching* rule that indicates how to check the value of the parameter provided at run time. Parameter matching allows us to associate a semantic type with the formal parameter of the operation. It specifies the set of information objects or attributes that are intended to be bound to the parameter. The matching rules are specified as an invariant on an information object relationship or as an attribute value of an information object. An information object is either specified directly or as a role played in an information relationship. These rules constrain the acceptable parameter more than the parameter type declaration. For example, while the parameter type may simply be integer, implying any integer will do, the parameter-matching rule may constrain this further to be a particular attribute on a particular object. While the previous discussion tacitly assumes that the parameter is a scalar, composite parameters (such as arrays, structures, and sets) can be used and each element of the composite can then be individually constrained by the parameter-matching rule. If a run-time parameter does not meet its parameter-matching rule, an exception will be raised and the appropriate error handling procedures will be invoked.

So far we have discussed interfaces from the point of view of a server, where the operations request the server to do something. In network management we frequently have the case when a client interface announces that some condition has been met and is not interested in what may occur as a result of this notification. Whereas a server is controlled by pre- and postconditions, the client is controlled by the *triggering condition,* which specifies the condition under which the notification will be emitted. The triggering condition is specified as if it were a precondition.

To summarize:

- Computational objects interact via interfaces, which are fully specified by their operations and behavior.

- A given computational object may have multiple interfaces, which can be instantiated from the same or different interface types.

- The most atomic interaction is the signal, and interfaces that support signals are called signal interfaces. Operations can be modeled using signals to express their relationship to the engineering infrastructure.

- Changes in the state of a system occurring as the result of operations at one interface may be viewed at other interfaces of the same object or at interfaces to other objects.

- Communication of state changes between computational objects may only occur via interactions through their bound computational interfaces.
- The transfer of continuous information flows are via stream interfaces.
- The decomposition of an application into a set of interacting computational objects establishes the limits for its potential distribution.

We note that typical management applications are built using the operational interface model, so we will focus only upon that in the remaining material.

4.2.3.3 Structure

Section 4.2.3.1 described the scope of the computational viewpoint and discussed the concepts needed to understand computational specifications. A complete computational viewpoint design requires the definition of computational objects as well as interfaces and operations. The definition of objects, as opposed to only interfaces, allows for the interaction between interfaces to be defined as well as the explicit definition of the number and types (client/server) of interfaces to be supported by a single object. This provides for a complete definition of the application related to the object. The definition of a single object with multiple interfaces precludes the need to deal with the replication of the state of a specific resource or other managed entity in multiple objects. If an object's interfaces need to be distributed, then the object may itself be decomposed into multiple computational objects.

The complete computational viewpoint design corresponding to an enterprise viewpoint community includes:

- *Computational operation design:* Specification of the operations required to support the enterprise actions specified for the system and to view data defined in the information specification.
- *Computational interface design:* Computational interfaces group closely related operations into a single interface. Operations are considered to be closely related when they have behavior influencing the common state of the object providing the interface. All of the operations specified in an interface type must be implemented by a single object instance on a single data processing node.
- *Computational object design:* Just as computational interfaces group closely related operations, computational objects group closely related interfaces. Again, closely related means that the interfaces have a high coupling as the common object state is affected by the interfaces. For

example, if one interface controls some value and another interface presents that value, it would be difficult to justify not providing these interfaces on a single object.

The system designer must take care when defining object types as collections of interface types. Indeed, when gathering interface types into a single object type instead of several objects, the designer limits possible distribution. To be more precise, computational object instances constitute the unit of distribution, which implies that a computational object instance and all the computational interface instances it supports are necessarily colocated. Gathering computational object and interface types is therefore not distribution neutral.

Having specified computational interfaces, we are in a position to say that we know about all the services that are offered, or invoked, within the system, without making any statement about how these services should be distributed over the processing elements of the system. The telecommunications world has decided that this level of detail is sufficient for standardization purposes and telecommunications standards do not specify computational objects, which are left to vendors as part of their implementations. We note that implementations involve internal interfaces and interactions that are clearly not subject to standardization.

The detailed structure of a computational viewpoint specification is provided in Appendix 4C. Just as for the other viewpoints, in order to write a computational specification it is necessary to have a language to write in and RM-ODP does not impose any constraints. However, RM-ODP does propose a candidate language called the ODP interface definition language (IDL), derived from the OMG specifications. The OMG forum, representing major software and telecommunication industry vendors interested in distributed computing, has demonstrated its willingness to design RM-ODP compliant platforms. The decision of ISO/ITU-T to include the definition of a candidate language at the RM-ODP computational viewpoint is exceptional (there is no proposal for the other viewpoint languages) and a reflection of pragmatic industry considerations. ODP IDL does not capture all of the computational viewpoint concepts. However it does provide constructs to define operational interface types by the set of operations that can be invoked through these interface types. Each operation is defined by its signature, that is, its name, its input and/or output parameters, and its exceptions. We note that the G.85x series of recommendations [5–8] did create a concrete syntax derived from GDMO-like templates, with semantics mainly taken from the ODP IDL (which also allows for accurate references between the computational and information viewpoints).

Mapping to Different Communications Domains

The point of specifying the management system in terms of interfaces and information objects is to allow for various communications infrastructures to be used in the final realization of the system. For example, CORBA is rapidly becoming the infrastructure of choice for distributed systems, and CMIP represents the current standard for TMN communications. Mapping between an ODP set of specifications and a communications infrastructure involves nothing more than identifying the constructs of the target infrastructure that support the concepts of the specification.

For the case of a CORBA-based infrastructure, as might be surmised, the computational interface specifications map relatively easily to the CORBA IDL. In particular, the <interface_reference> specification type is directly supported in CORBA. The mapping therefore involves rewriting the interface definitions in terms of CORBA IDL and allocating CORBA objects to support those interfaces. The actual protocol used between CORBA objects is well hidden from the IDL specifications, so we do not have to take that into account at this stage.

Mapping to the CMIP infrastructure, introduced in Chapter 2 (involving GDMO and ASN.1), is more complicated. In providing communications between a manager and managed system, as discussed in Chapter 2, the messages carried by CMIP are defined using the GDMO notation standardized in X.722 [9]. Specifically, GDMO defines the structure of information objects to be carried by CMIP, while the data types of the message are specified using Abstract Syntax Notation 1 (ASN.1) standardized by X.208 [10]. There is a fundamental difference between computational interfaces and GDMO objects. The computational viewpoint makes information object attribute data available by defining an interface, whereas GDMO defines an object for the same purpose. Mapping computational interfaces to CMIP therefore involves creating GDMO objects for each interface specified in the computational viewpoint and creating ASN.1 data types to support the information entities (e.g., objects and relationships) specified for the system. While interfaces may inherit the properties of other interfaces, GDMO objects must always inherit from some suitable "top" object. Mapping to GDMO therefore involves creating a suitable inheritance tree for the GDMO objects representing the system interfaces. Finally, the <interface_reference> specification type is not directly supported in GDMO. It is therefore necessary to use the <object instance type> (Object-Identifier), which is much less precise than the object interface reference in the original IDL. This will be further discussed in Chapter 7.

Multiprotocol Communications Domain Computational Viewpoint

The computational viewpoint can be divided into two parts: a part that is mostly independent of the underlying engineering environment or communications domain and a part that is more closely linked to the communications domain used in the engineering viewpoint (e.g., CMIP/OSI and CORBA), as shown in Figure 4.9.

The goal of the separation into parts is to preserve the behavior (dynamic schemas) as specified in the information viewpoint while enabling different engineering implementations using different communications infrastructures. As we have seen, there is more than one possible computational viewpoint corresponding to a given information viewpoint, so the amount of protocol independence depends upon the extent to which a given engineering protocol supports the concepts of the particular computational viewpoint. As an example, if the protocol chosen does not support the concept of an object as defined by the object-oriented paradigm introduced in Chapter 2 (e.g., SNMP), then a different computational viewpoint is needed and it is essential to design a representation of an object in the infrastructure. This involves yet more work.

Mappings to Communications Domain-Dependent Computational Viewpoint Specification

Computational templates defined using the G.851 computational viewpoint language [5] must be mapped to templates for the communication domain used in the engineering realization. A subset of ASN.1 abstract syntax is used in this computational viewpoint language to define the syntax of the computational operation parameters. This abstract syntax representation and the computational operations must be mapped to the particular syntax and protocol for

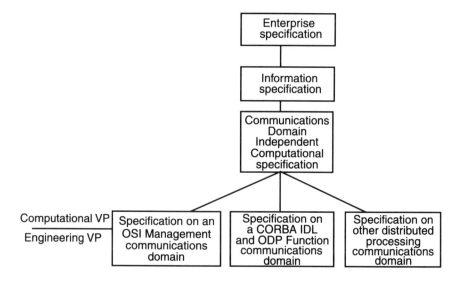

Figure 4.9 Alternative communications domain specifications.

the communication domain used in the engineering realization. This process results in a computational specification that is specific for each intended communication domain.

4.2.4 Engineering Viewpoint

The engineering viewpoint focuses on describing the infrastructure aspects to support distribution, that is, what components have to be present to enable communication between remote computational interfaces.

4.2.4.1 Scope

RM-ODP defines modeling concepts that are powerful enough to take into account any situation since the basic assumption of RM-ODP is maximal heterogeneity, meaning, for example, that the implementation of the system can be:

- *Widely distributed:* all implementation objects are distant from each other (one object per processing node).
- *Highly heterogeneous:* in terms of degree of automation, execution platforms, and communications infrastructures.
  - The service provided by some implementation objects might be handmade by a human agent whereas the service offered by other implementation objects might be completely automated.
  - When implementation objects are executed on computers, they can be implemented using different operating systems, programming languages (object-oriented or not), and database management systems, for example.
  - The various platforms that are involved might use different communication protocols.

The purpose of the engineering specification is to describe the whole infrastructure that is required to enable the basic engineering objects (BEO) to communicate without having to concern themselves with distribution issues. Usually, a BEO is the equivalent of a computational object instance, from the engineering point of view. This means that a BEO is an implementation application-level object. RM-ODP also defines the concept of transparency objects that are implementation infrastructure-level objects that enable communication between remote BEOs. Since this book does not intend to provide guidance on developing a platform, the modeling concepts introduced to support this (e.g., stub and binder) will not be presented here in detail. More detailed information on this can be obtained from RM-ODP Part 3 [3].

4.2.4.2 Modeling Concepts

As introduced previously, two kinds of engineering object types are defined in RM-ODP:

- *BEO type:* Each computational object type, together with its interface type, is associated with a so-called basic engineering object type. A BEO constitutes the engineering view of a computational object.

- *Transparency object type:* When distribution and/or heterogeneity is to be supported, such object types must be defined in the engineering specification. RM-ODP defines several kinds of transparency object types. Transparency engineering object types called stubs, among other things, encode and decode (marshal/unmarshal) protocol data units corresponding to potential requests and responses exchanged between basic engineering objects. Other types of transparency engineering objects include binders, whose aim is to provide mechanisms to ensure the quality of intercomputational interfaces communications, and interceptors that are necessary to enable protocol conversion when two potentially interacting computational interfaces have stubs that support different communication protocols. As we will see in Chapter 6, in the functional modeling terms of Chapter 3, an interceptor can be seen as a pair of adaptation functions connected back-to-back whose input is protocol 1 data units and output is protocol 2 data units.

4.2.5 Technology Viewpoint

The technology viewpoint focuses upon describing the implementation of the enterprise in terms of actual infrastructure components. For example, if the enterprise is to be automated, both the hardware and software architectures need to be described in terms of particular computing engines (e.g., PCs, Work Stations), operating systems, database management systems, programming languages, and communication protocols, for example. RM-ODP places very little emphasis on this viewpoint. As a consequence, any distributed system specification can claim conformance to the RM-ODP technology viewpoint.

4.3 Summary

Open distributed processing provides a modeling technique that supports the capability to specify network behaviors without constraining the physical distribution of those behaviors and to specify logical object-oriented software architectures independent of their physical distribution. This "future-proofs" the specification by making it independent of evolving distribution capabilities and frees manufacturers to distribute the functionality in the ways they choose, to best suit their customers' requirements and optimize their cost of development. Applied to network management, it supports distribution and object-oriented design to circumvent the processing and communications limitations of the current environment and to provide the fine-grained modularity needed to support a multivendor environment and migration of functionality. Its utility is not limited to what we refer to as a "centralized" or "hybrid" environment. (In fact, when we use the term "centralized" to characterize current network management applications, we are actually referring, in most cases, to a distributed architecture in which there is a "master-slave" relationship between the "centralized" network management system and subordinate computing platforms residing either within or near the network elements.) In other words, the ODP reference model simply offers a more generic specification tool than those that depend on prior knowledge of the physical distribution of data and functionality.

This chapter dealt with the various viewpoints in relative isolation. As indicated in the brief discussion on computational objects, there is a coupling between the viewpoints. The example in Chapter 5 should help clarify how viewpoints are related and how design decisions are made when creating the next viewpoint description.

References

[1]   ISO/IEC 10746-1 / ITU-T Rec. X.901 "Information Technology, Open Distributed Processing, Reference Model: Overview," 1997.

[2]   ISO/IEC 10746-2 / ITU-T Rec. X.902 "Information Technology, Open Distributed Processing, Reference Model: Foundations," 1995.

[3]   ISO/IEC 10746-3 / ITU-T Rec. X.903 "Information Technology, Open-Distributed Processing, Reference Model: Architecture," 1995.

[4]   ISO/IEC 10746-4 / ITU-T Rec. X.904 "Information Technology, Open Distributed Processing, Reference Model: Architectural Semantics," 1997.

[5]   ITU-T Rec. G.851-01 "Management of the Transport Network, Application of the RM-ODP Framework," 1996.

[6]   ITU-T Rec. G.852-01 "Management of the Transport Network, Enterprise Viewpoint for Simple Subnetwork Connection Management," 1996.

[7]   ITU-T Rec. G.853-01 "Common Elements of the Information Viewpoint for the Management of a Transport Network," 1996.

[8]   ITU-T Rec. G.854-01 "Management of the Transport Network, Computational Interfaces for Basic Transport Network Model," 1996.

[9] ITU-T Rec. X.722/ISO/IEC 10165-4: 1992, "Information Technology, Open Systems Interconnection, Structure of Management Information, Part 4, Guidelines for the Definition of Managed Objects."

[10] ITU-T Rec. X.208, "Open Systems Interconnection Model and Notation—Specification of Abstract Syntax Notation One (ASN.1)," 1988.

Appendix 4A
Enterprise Viewpoint Structure

4A.1 Informal Definition of the Enterprise Template

This section introduces in an informal way the templates that will be filled when specifying an enterprise community. The reference to X in the templates indicates the number of the section in a document in which the community specification is to be inserted.

X COMMUNITY <community_label> "Name"

X.1 PURPOSE

This clause introduces the purpose of the community.

X.2 ROLE

This clause introduces the set of roles that are used in the community.

X.3 POLICY

This clause introduces the set of policy statements that are applicable for the entire community, each one being introduced with a label and a qualifier stating if the statement is either a PERMISSION, an OBLIGATION, or a PROHIBITION. Each clause has to mention the role concerned by this constraint.

X.4 ACTION

This clause introduces the set of actions that is made available in the community.

X.4.1 "Action Name"

This subclause introduces the semantic of each action by a definition and a set of policy statements, each one being introduced with a label and a qualifier stating if the statement is either a PERMISSION, or an OBLIGATION, or a PROHIBITION.

X.5 ACTIVITY

This clause introduces the set of activity that is made available in the community.

X.5.1 "Activity Name"

This subclause introduces the semantic of each activity by a definition, the action graphs corresponding to the activity, and a set of policy statements, each one being introduced with a label and a qualifier stating if the statement is either a PERMISSION, or an OBLIGATION, or a PROHIBITION.

4A.2 Formal Definition of the Enterprise Template

This section provides a formal description of the enterprise description in BNF.
Note - extensions to BNF for this specification
   [...] - optional item, 0 or 1 occurrences
   * list item 1 or more occurrences
   -- remainder of line is a comment as per ASN.1

```
<community_template> ::=
   "COMMUNITY" <label> <name>
   "PURPOSE" <community_definition>
   "ROLE" <role_definition>*
   "POLICY" <policy_definition>*
```

"ACTION" <action_description>*
"ACTIVITY" <activity_definition>*

["WITH ACTION GRAPH"
    "Start"
    <action_label>*
    "End"]
"CONTRACTS" <label_string>
<action_description> ::=   <label> <label_string>
                          "ACTION POLICY" ":" <policy_definitions>
<action_label> ::=   <label>
<activity_definition> ::= <label> <label_string>
                          "ACTIVITY POLICY" <policy_definitions>
<activity_definitions> ::=   "None" | <activity_definition>*
<community_definition> ::=   <label_string>
<doc_heading> ::=   <text>
                    -- doc_heading represents a heading when the template
                    -- is printed. It is publishing system specific and would
                    -- be considered White space in an implementation
<label> ::=   "[A-Za-z][-A-Za-z0-9.]*"
<label_string> ::= -- quoted string | label
<name> ::=   <label>
<policy_definitions> ::=   "None" | <policy_definition>*
<policy_definition> ::=   { "PERMISSION" | "OBLIGATION" | "PROHIBITION" |
                          "EXCEPTION" } <label> <label_string>
<role_definition> ::= <role_label> <label_string>
<role_label> ::=   <label>

Appendix 4B
Information Viewpoint Structure

4B.1   Introduction

The information viewpoint section is structured into the following subsections:

- A list of information entities which have been defined in other specifications and are referenced by the corresponding enterprise community,
  including other information entities from other specifications for inheritance purposes. When information entities are imported into an information specification in this way, all other related specifications relevant to
  this entity are also imported (e.g., informal, semi-formal, and formal
  specifications are imported when one of these specifications is listed).

- Diagrams of information object and relationship types as required to provide readability:
  —contains role-relationship diagrams and class-role diagrams;
  —inheritance diagrams;
  —relationship diagrams.

- Information Object Classes
  —defines the object classes.

- Information Relationships
  —defines the relations in which the information objects participate.

- Static Schema Definitions
  —defines the global and generic system states, described in schema.

- Schema Transition Definitions (Dynamic Schemas)
  —labels particular transitions between compound states, described by static schema.

- Attribute Type Definitions
  —definition of the attributes for the concerned service.

4B.2 Informal Definition of the Information Template

This section introduces in an informal way the templates that will be filled when specifying an information viewpoint specification. The reference to X in the templates indicates the number of the section in a document in which the specification is to be inserted.

X.1 Information Object Classes

This clause introduces the set of information object classes that are part of the information viewpoint specification.

X.1.1 "Information Object Class Name"

This subclause specifies each information object class and may be structured into four or five parts:

• The first one, introduced by the keyword DEFINITION, contains a definition of the information object class.
• The second one, introduced by the keyword ATTRIBUTES, lists the attributes of the information object class by inclusion of a reference, which can be local if the attribute is locally defined or global if the attribute is imported from another service.
• The third one, introduced by the keyword INVARIANT, describes the permitted states for an instance of the information object class.
• The fourth part, introduced by the keyword POTENTIAL RELA-TIONSHIPS, expresses the possible relationships in which the information object or its subclasses can participate.
• The last part, introduced by the keyword RELATIONSHIPS, expresses the relationships in which the information object or its subclasses have to participate.

X.2   Information Relationship Class

This clause introduces the set of information relationship classes that are part of the information viewpoint specification.

X.2.1   "Information Relationship Name"

This subclause specifies each information relationship class and may be structured into four or five parts:

—The first one, introduced by the keyword DEFINITION, contains a definition of the relationship class.
—The second one, introduced by the keyword ROLE, describes the roles of the relationship class.
—The third one, introduced by the keyword INVARIANT, gives the invariants applicable to the information objects playing the roles of the relationship during the lifetime of this relationship.
—The last part, introduced by the keyword TRANSITION, expresses the restriction on the state transitions of the information object playing the roles of the relationship.

X.3   Static Schema

This clause introduces the set of static schemas that are part of the information viewpoint specification.

X.3.1 "Static Schema Name"

This subclause specifies each static schema relevant for the specification and may be structured into three parts:

—The first one, introduced by the keyword DEFINITION, contains a definition of the static schema.
—The second one, introduced by the keyword ROLE, describes the roles involved in the schema.
—The third one, introduced by the keyword INVARIANT, gives the invariants applicable to the schema.

X.4 Dynamic Schemas

This clause introduces the set of dynamic schemas that are part of the information viewpoint specification.

X.4.1 "Dynamic Schema Name"

This subclause specifies each dynamic schema relevant for the specification and may be structured into three parts:

—The first one, introduced by the keyword DEFINITION, contains a definition of the dynamic schema.
—The second one, introduced by the keyword PRE_CONDITION, indicates, either by reference or explicitly, the static schema that acts as pre condition.
—The third one, introduced by the keyword POST_CONDITION, indicates, either by reference or explicitly, the static schema that acts as post condition.

X.5 Attributes

This clause introduces the set of attributes that are part of the information viewpoint specification.

X.5.1 "Attribute Name"

This subclause specifies each attribute relevant for the specification and is structured in four parts:

—The first one, introduced by the keyword DEFINITION, contains a definition of the attribute.

—The second one, introduced by the keyword STATE, lists all the values that an attribute can have, with associated semantics.

—The third one, introduced by the keyword INVARIANT, lists the invariants valid for that attribute, if any.

—The fourth one, introduced by the keyword TRANSITION, lists all the possible transitions between the states of the attribute. Defining a transition table can do this.

Appendix 4C
Computational Viewpoint Structure

4C.1 Introduction

The computational viewpoint section is structured into the following subsections:

- A list of computational object classes;
- A list of computational interface classes;
- A list of computational operations.

4C.2 Informal Definition of the Computational Template

This section introduces in an informal way the templates that will be filled when specifying a computational viewpoint specification. The reference to X in the templates indicates the number of the section in a document in which the specification is to be inserted.

X.1 Computational Object Classes

This clause introduces the set of computational object classes that are part of the computational viewpoint specification.

X.1.1 "Computational Object Class Name"

This subclause specifies each computational object class and may be structured into four parts:

—The first one, introduced by the keyword COMPUTATIONAL_OBJECT_CLASS, gives the name of the computational object class.

—The second one, introduced by the keyword SERVER_INTERFACES, gives the bound server interfaces to the computational object class.

—The third one, introduced by the keyword CLIENT_INTERFACES, gives the bound client interfaces to the computational object class.

—The fourth one, introduced by the keyword BEHAVIOUR, may be used to indicate the behavior of the computational object class.

X.2 Computational Interface

This clause introduces the set of computational interfaces that are part of the computational viewpoint specification.

X.2.1 "Interface Class Name"

This subclause specifies each interface class and may be structured into three parts:

—The first one, introduced by the keyword COMPUTATIONAL_ INTERFACE, gives the name of the computational interface.

—The second one, introduced by the keyword DERIVED FROM, indicates inheritance characteristics.

—The third one, introduced by the keyword OPERATION, gives the list of the operations available on that interface.

X.3 Operations

This clause introduces the set of operations that are part of the computational viewpoint specification.

X.3.1 "Operation Name"

This subclause specifies each interface class and may be structured into five parts:

—The first one, introduced by the keyword OPERATION, gives the name of the operation.

—The second one, introduced by the keyword INPUT_PARAMETERS, indicates the name and the syntax of each input parameter in the operation signature. The ASN.1 notation is used for the parameter syntax.

—The third one, introduced by the keyword OUTPUT_PARAMETERS, indicates the name and the syntax of each output parameter in the operation signature. The ASN.1 notation is used for the parameter syntax.

—The fourth one, introduced by the keyword RAISED_EXCEPTIONS, indicates the name and the syntax of each exception in the operation signature. The ASN.1 notation is used for the parameter syntax.

—The fifth one, introduced by the keyword BEHAVIOUR, specifies the behavior of the operation. This behavior is itself structured in the following subparts:

◊ the first one, introduced by the keyword INFORMAL, is used to describe it in an informal way using the natural language

◊ the second one, introduced by the keyword SEMI INFORMAL, will be used to more formally relate the signature to an information viewpoint content by providing first a matching clause (introduced by PARAMETER-MATCHING keyword), secondly by specifying the precondition semantics (introduced by PRE_CONDITIONS keyword), thirdly by specifying the postcondition semantics (introduced by POST_CONDITIONS keyword), and finally by specifying the exception semantics (introduced by EXCEPTIONS keyword). If the operation is a notification, the pre- and postcondition subclauses are related by a triggering condition clause (introduced by TRIGGERING_CONDITIONS keyword).

4C.2 Informal Definition of the Computational Templates

This section defines the computational templates using BNF.

4C.2.1 Object Class Template

```
<computational_object_template> ::= <computational_object_header>
                        "{"<computational_object_body> "}"
<computational_object_header> ::= "COMPUTATIONAL_OBJECT_CLASS
<object_name>
<object_name> ::=   <identifier>
<computational_object_body> ::= [<server_interface_definitions>]
                        [<client_interface_definitions>]
                        [<behaviour_definition>]
```

```
<server_interface_definitions> ::= "SERVER_INTERFACES"
{<server_interface_label> ";" }*
<server_interface_label> ::= <label_reference>
<client interface_definitions> ::=   "CLIENT_INTERFACES"
{<client_interface_label> ";" }*
<client_interface_label> ::= <label_reference>
<behaviour_definition> ::= "BEHAVIOUR" {<text_delimiter> <string_literal>
      <text_delimiter>
                            | <string_literal> }";"
<identifier> ::= [a-zA-Z][-a-zA-Z0-9_:.]*
   -- This accepts identifiers as starting with a letter and containing
   -- letters, digits, underscores, hyphens, colons and points.
<text_delimiter> ::= ! | " | # | $ | % | ^ | & | * | ' | ` | ~ | ? | @ | \
```

Note: If a text_delimiter is used, the same character shall be used at the start and end of the string, and whenever that text_delimiter character appears in the body of the text string, it shall be replaced by two occurrences of that character. If a text_delimiter character is not used, then the text string shall not contain any punctuation character that is a valid successor to the text string in the BEHAVIOUR template (i.e. ";").

4C.2.2 Interface Template

```
<computational_interface_template> ::= <computational_interface_header>
                            "{" <computational_interface_body>
"}"
<computational_interface_header> ::=   "COMPUTATIONAL_INTERFACE
<interface_name>
<interface_name> ::= <identifier>
<computational_interface_body> ::= ["DERIVED FROM" <interface_label>]
                            <operation_definitions>
                            [<behaviour_definition>]
<interface_label> ::= <server_interface_label> | <client_interface_label>
<operation_definitions> ::= "OPERATION" {<operation_label> ";"}*
<operation_label> ::= <label_reference>
```

4C.2.3 Operation Template

```
<operation_template> ::=   <operation_header> "{" <operation_body> "}"
<operation_header> ::=   "OPERATION" <operation_name>
<operation_name> ::= <identifier>
<operation_body> ::= [{INPUT_PARAMETERS" [{<param_label> ":" <syntax_label> ";"}* ]}]
                    [{OUTPUT_PARAMETERS"        [{<param_label>          ":"
                    <syntax_label>";"}*]}]
```

```
                    [{RAISED_EXCEPTIONS"                        [{<exception_label>":"
                    <syntax_label>";"}*]}]
                          [<opn_behaviour_definition> ";"]
<param_label> ::=   <identifier>
<exeption_label> ::=   <identifier>
<syntax_label> ::=   <primitive_asn1_type_name> | <module_name> "::"


<production_name> |<type_production> | <production_name>
<production_name> ::=   <identifier>
<module_name> ::= <identifier>
<primitive_asn1_type_name> ::=   <identifier>
<type_production> ::=     <type> "::=" <comp_type_def>
<comp_type_def> ::= <interface_reference> | <singleASN1typedef> -- Imported from X.208
<interface_reference> ::=   "("<interface_type_name>")"
<interface_type_name> ::=   <identifier>
<opn_behaviour_definition> ::= "BEHAVIOUR"
             "INFORMAL" [{<text_delimiter> <string_literal> <text_delimiter>
             | <string_literal> ";"}]
             "SEMI_FORMAL"
                   {"PARAMETER_MATCHING"
                   {<param_label> ["ELEMENTS"]":"
                                    <parameter_matching_expression> ";"}*}
                   {[{"PRE_CONDITIONS" [{<schema_label> | {<text_delimiter>
                                    <string_literal> <text_delimiter>}";"}]}]
                   [{"POST_CONDITIONS" [{<schema_label> | {<text_delimiter>
                                    <string_literal> <text_delimiter>}";"}]}]} |
                   [{"TRIGGERING_CONDITIONS"
                         [{<transition_label> | <text_string> |
                         <state_label> "TRANSITION_TO" <state_label> ";" }]}]
                   [{"EXCEPTIONS"
                   [{"IF" <exception_invariant_label> "NOT_VERIFIED"
                         "RAISE_EXCEPTION" <exception_label> ";"}*]}]
<schema_label> ::= <label_reference>
<invariant_label> ::= <label_reference>
<exception_invariant_label> ::=   {"PRE_CONDITION" <label_reference>}
                                | {"POST_CONDITION" <label_reference>}
<transition_label> ::= <label_reference>
<text_string> ::=   <string_literal>
<string_literal> ::= [a-zA-Z][-a-zA-Z0-9_:.]*
<string_literal> ::= [a-zA-Z][-a-zA-Z0-9_:.]*
<state_label> ::=   <label_reference>
                   <parameter_matching_expression> ::=
```

5

Creation of a Management Service Specification

5.1 Introduction

In the previous chapter, we introduced the open distributed processing framework of abstractions and basic concepts associated with each of the viewpoints. In this chapter we will apply these concepts to the telecommunications domain, informally examining the configuration management service. We will show that this service can be decomposed into a set of more elementary services and describe how to specify one of the simplest: topology management. Specifically, we will provide a description of the enterprise, information, and computational viewpoints for this elementary service together with an explanation of how the viewpoint specifications are mutually developed.

5.2 Architecture Specification Process

While Chapter 4 describes how to specify a service in terms of RM-ODP concepts augmented with telecommunications domain application methods, we have not applied this approach to specify any particular service. We will now explain how this may be done. Let us begin by considering the problem a team of specifiers faced a number of years ago. Specifically, "We have a transport network to manage, in particular, an SDH transport network. How do we accomplish this?" In order to provide a potential client with an end-to-end connection across the network, it was clear that a subnetwork connection management service was needed. Thus, it was decided that specifying the SDH path setup service, which is of great interest to transmission network operators, would be a good place to start. It rapidly became obvious, however, that providing an exhaustive management service specification was far too complex a task to accomplish as a one-stage process. In particular, having to consider the associated set of problem domain resources, interactions, and dependencies at the same time as addressing solution domain management service definition, policies, and actions formed a considerable barrier to progress. This barrier was surmounted when the team introduced a two-stage process for specifying transport network management services:

1. Identify and model the set of static resources that are manageable in the transport domain (i.e., the management resources introduced in Chapter 3), independent of the management services that may be using them.

2. Establish an approach based on defining a set of elementary services (atomic), which manipulate these management resources, that will then cooperatively interact to provide the desired management service.

As introduced in Chapter 3, transport entities can be considered as management resources, where a management resource represents the visibility a manager, or a component for the purposes of management, has into a transport network. In fact, it is possible to provide a catalogue of enterprise resources essential to the support of transport network-level management services [1]. The behavior of these enterprise resources can be specified using enterprise viewpoint constructs written in textual style. An information specification for these enterprise resources may also be generated using information viewpoint constructs introduced in Chapter 4, and it is similarly possible to provide a catalogue of information objects [2]. A range of management services (e.g., configuration, monitoring, and accounting) may utilize these information objects.

As discussed earlier, the approach taken is to decompose a management service into a set of cooperatively interacting elementary services, each of which can be more easily specified. However, how do we decide upon an appropriate decomposition? In fact, the choice is not arbitrary but is based upon appropriate utilization of management service domain engineering expertise, that is, the years of experience and common practices of network operators worldwide. Capturing the dynamic requirements of a specific management service (e.g., subnetwork connection management) as an enterprise specification requires an understanding of the associated actions and activities, as discussed in Chapter 4. Some understanding of management service actions and activities, in addition to pragmatic constraints, may also be obtained from telecommunications operator business process models (e.g., those defined within the TeleManagement Forum[1]), which reflect a range of relevant business concerns [3]. Thereafter, the corresponding information, computational, and engineering viewpoint descriptions for the service can be developed. Engineering and technology aspects are additionally affected by such factors as infrastructure constraints (e.g., execution, communication). The overall management service specification process is illustrated in Figure 5.1.

Cataloging sets of services (e.g., in standards) specified according to the preceding process facilitates rapid deployment of telecommunications operator services.

5.3 Application to Subnetwork Connection Management Service

In Chapter 3 we discussed the topological elements of any network, while in Section 3.3.2 we discussed the management resource view of those elements.

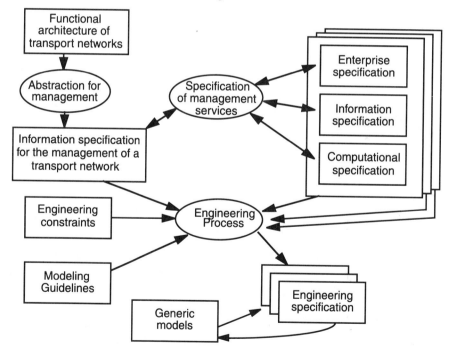

Figure 5.1 Management service specification process.

---

1. The TeleManagement Forum (formerly called the Network Management Forum) is a non-profit, global organization that provides the telecom industry with leadership on the most effective ways to streamline the management of communications networks and services. More details may be found at URL http://www.tmforum.org/.

These resources constitute the "domain knowledge" of the domain of network management, which is the body of knowledge available to the specifier. In order to provide some context, we will briefly and informally sketch the domain of network configuration management before turning our attention to the subnetwork connection management service in particular. We will be discussing resources available to a manager and the services that manipulate those resources using the terms introduced in Chapter 3.

Within the network management domain, perhaps the most important service to consider is that of network configuration management, the ultimate goal of which is to provide a client with their desired connections. The two most important elements of the topology model are subnetworks and links, so the most important elements of this service (see Figure 5.2 and Chapter 3) are subnetwork connection management and link connection management. Subnetwork connection management is responsible for making a connection across a subnetwork, while link connection management is responsible for assigning a link connection between subnetworks in order to enable a subnetwork connection to be completed. Of course, before connections can be established, resources must first be made available; this is accomplished by the topology management service, which is responsible for the creation of the

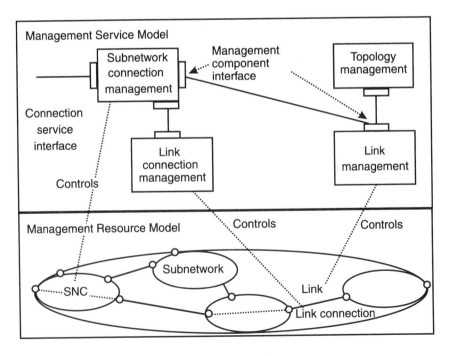

Figure 5.2 Subnetwork connection management service.

subnetwork and link resources required to provide potential transport capacity. Once a link has been created, link management takes capacity from a server and makes it available to the link (i.e., manages link capacity in terms of available bandwidth by adding or reducing capacity). The actual capacity of the link is provided by server layer trails that are controlled by the trail management service (not illustrated in the figure), which establishes or releases a trail between a specified set of endpoints and/or access groups at the boundary of a layer network. Trail management also controls trail termination points (TTPs) and their association with subnetworks, access groups, or CTPs. Finally, adaptation management deals with the relationships between the capacities of the client link and the server trail. While this description implies a time ordering of operations, the individual services are only constrained by their preconditions. Link management, for example, could be dynamically invoked in order to provide additional capacity when a particular link became overloaded.

These management services are generic in that they are described independently of the underlying transport technology. Applications can be differentiated by whether the implementation technology allows for a preprovisioning of the different capacities, as in SDH/SONET or OTN, or is more dynamic, as in ATM or IP. Chapter 9 will provide a concrete example dealing with the configuration of a multitechnology transport network that makes use of these services.

In this chapter we will concentrate on the topology management service, which is used to create and delete the objects representing the resources that compose a layer network domain, regardless of the type of the network. This service is small enough to treat in its entirety but complex enough to illustrate all the major principles. While this example may feel both trivial and complex, the feeling of triviality is caused by the simplification created by the layered model of Chapter 3. Feelings of complexity arise from the explicit nature of the specification technique, where anything of consequence is explicitly captured. Incomplete specifications with many implicit details should not be confused with simple specifications since, in general, the true complexity only becomes apparent too late. Of course, we must be careful to only capture the complexity of the problem rather than the accidental complexity of a particular solution to the problem.

The observant reader may notice that this service constructs abstract networks in that the subnetworks and links created have no obvious physical actuality. As we have seen in Chapter 3, a layer network can be recursively decomposed until the subnetworks represent physical equipment (the matrix) and the links represent physical trails provided over some physical medium. In the case of the topology management service, we recursively construct larger networks by assembling them from matrices and physical link connections (e.g., optical fibers and copper wire). In the limit, the topology service binds the objects created to these actual physical resources that provide physical transport functions.

5.3.1 Overview of Viewpoint Relationships

To aid in the understanding of the material that follows, in this section we will provide a "quick reference guide" to key interviewpoint relationships. As illustrated in Figure 5.3, enterprise roles map onto information objects, enterprise policies map onto information attributes and relationships, and enterprise actions map onto computational operations. Relationships are used to represent the various kinds of constraints needed in all viewpoints (e.g., enterprise policies and computational parameter-matching rules). Computational parameters use information objects and attributes, whereas computational constraints (expressed in terms of preconditions and parameter-matching) use information relationships.

5.3.2 Management Resource Model

A management resource is an entity that can play a static role in some enterprise community. Management resources are specified by a definition, which clearly describes the purpose of the resource, and a list of properties that constrain the resource. The subset of resources from the library of management resources relevant to the topology management service (previously introduced in Chapter 3, Section 3.3.2) are:

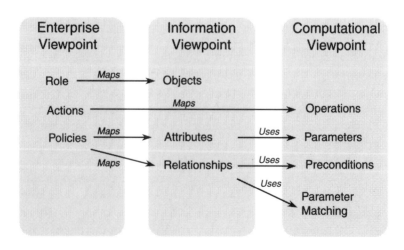

Figure 5.3 Interviewpoint relationships.

- *Layer network domain*—the set of the managed resources with a given characteristic information.
- *Subnetwork*—used to effect routing of a specific signal, which may be characterized by a set of related extremities.
- *Link*—represents the transfer capacity between two subnetworks, or one subnetwork and one access group.
- *Link end*—represents the extremity of a link.
- *Access group*—a set of trail termination points providing access to the layer network.
- *TTP*—represents a point terminating a trail.
- *Link connection*—transfers information between two points of two subnetworks in a fixed manner.
- *Subnetwork connection*—transfers information across a subnetwork in a flexible manner.
- *CTP*—represents the extremity of a link connection.

5.3.3 Management Resource Information Specification

Each management resource is mapped onto an information object class, which is usually identified by the same name as the management resource. As described in Appendix 4B, information objects are specified by their definitions, attributes, and potential relationships. Management resource properties, which correspond to policies that apply to a single role,[2] map into the attribute and relationship clauses in the information specification. These information object attributes are generic in that they represent information useful to all management services. Thus, defining attributes in the resource model that will only be used by one specific service is a practice that should be avoided. The *potential relationship* clause allows us to list all the relationships in which the defined object might be involved. This should not be taken as inferring that all these relationships will be needed to support any particular service that might use the resources later. The service designer will select the relevant subset of relationships needed to support their particular service.

We will provide two examples illustrating the derivation of information specifications from management resources. The examples are given in the form of two columns, with the left providing the management resource description and the right providing the information specification. Corresponding information between the columns is placed side by side to highlight the management

---

2. The enterprise "properties" keyword clause is used here as a keyword though it has not been defined as such in the grammar of Appendix 4A.

resource description from which the information specification constructs are derived. The derivation of the information object "layer network domain" is provided in Table 5.1, and the derivation of the information object "subnetwork" is provided in Table 5.2.

In the preceding example, we used a number of relationships. In Table 5.3, we will use the relationship <snIsPartitionedBySn> as a vehicle to illustrate how an information relationship may be specified.

5.4 Topology Management—Enterprise Viewpoint

Having given a flavor of how information objects are derived from management resources, we will describe the topology management community purpose,

Table 5.1
Derivation of Information Object Layer Network Domain

Management Resource Enterprise Role	Management Resource Information Object
layer network domain	*LayerNetworkDomain* This information concept is related to the enterprise role layer network domain (Lnd).
Definition A layer network domain is a transport administrative domain in which all resources pertain to the same G.805 layer. It represents the network entities that together provide communication services with one signal identification.	DEFINITION "A layer network domain object represents a transport administrative domain in which all resources pertain to the same G.805 layer. The layerNetworkDomain information object type is a subtype of the networkInformationTop information object type."
Properties RELATIONS: A layer network domain may be related to other layer network domains as a server or client. The corresponding signal identifications shall have an adaptation relationship. For example, an SDH VC-4 layer network domain may serve SDH VC-11, VC-12, VC-3, or ATM VP layer network domains.	ATTRIBUTE signalIdentification "The signalIdentification describes the signal that is transferred across the layer network domain." POTENTIAL RELATIONSHIPS <layerNetworkDomainCanServeLnds> <layerNetworkDomainIsMadeOf>

Table 5.2
Derivation of Subnetwork Information Object

Management Resource Enterprise Role	Management Resource Information Object
subnetwork	*Subnetwork* This information concept is related to the enterprise role subnetwork.
Definition A subnetwork represents a topological component used to effect routing of a specific signal identification (see G.805: 1995 definition). It may be characterized by a set of related extremities, which can be link connections, CTPs, or TTPs and which shall be connectable.	DEFINITION "A subnetwork information object represents 'a topological component used to effect routing of a specific signal identification.'"
Properties SIGNAL IDENTIFICATION: A subnetwork carries a signal identification. The signal identification will be defined in the technology specific extensions.	ATTRIBUTE signalIdentification "The subnetwork carries a specific format. The specific format will be defined in the technology specific extensions."
COMPOSITION: A subnetwork may be composed of smaller subnetworks that are part of its decomposition, due to partitioning. Decomposition of subnetworks may be recursive; that is component subnetworks may be themselves composed of inner subnetworks, and so on. The smallest component subnetwork is associated with a G.805 matrix.	POTENTIAL RELATIONSHIPS <snIsPartitionedBySn> <snIsPartitionedByLinks>
COMPOSITION_CONSTRAINT: The component subnetworks must have the same signal identification as the composite one.	
CONNECTIVITY: A subnetwork may be crossed by subnetwork connections. At some point in time, the number of subnetwork connections crossing a given subnetwork may be zero.	<subnetworkHasSubnetworkConnections>

Table 5.2 *(continued)*

Management Resource Enterprise Role	Management Resource Information Object
RELATED_EXTREMITIES: A subnetwork can be defined by a set of related extremities (e.g., CTP, TTP, link connection). This set represents the potential connectivity of the subnetwork.	<subnetworkIsDelimitedBy> <subnetworkIsDelimitedBySnTpPools> <linkBinds> <linkConnectionIsTerminatedByTopological Entities> <linkEndIsBoundTo>
PROTECTED SUBNETWORK: A subnetwork may be considered as protected if all its subnetwork connections are protected.	

roles, and actions, focusing upon those related to subnetwork creation and deletion. As pointed out in Chapter 4, the enterprise viewpoint is the viewpoint that captures the requirements of the enterprise. In this section, we will focus our discussion upon how topology management requirements are captured in an enterprise specification versus elaborating on the requirements themselves.

The purpose of the topology management community is to manage the resources of the pertinent layer network domain. All resources manipulated by community actions must pertain to that community and are governed by the architectural interconnection rules described in Chapter 3. The service the community provides is to enable the creation and deletion of associated layer network domain resources (i.e., the subnetworks, links, link ends, and access groups) as well as to provide a set of reporting actions that indicate when community resource creation and deletion is taking place. Note that when we talk about resources, we usually mean a representation of a resource in some management system, since it is difficult to conceive of a management system physically creating a physical fiber link connection or physical equipment. In summary, the topology management service is realized as a result of actions to create, delete, or otherwise modify the resources that make up a transport network. The association and disassociation of TTPs and CTPs with the related subnetworks and access groups, and the partitioning of subnetworks and links, are managed by other services and are not explicitly addressed within this community.

5.4.1   Roles

The passive roles supporting the topology management service are those of the management resources detailed in Section 5.3.2, whose properties have become

Table 5.3
Description of Information Relationship

snlsPartitionedBySn Relationship	Description
This relationship type is related to the following enterprise entity: :<"ITU-T Draft Rec. G.852-02", COMMUNITY:tem, ROLE:sub-network, PROPERTY :composition>.	This is a reference to the enterprise document in which it is defined.
DEFINITION "The snlsPartitionedBySn relationship class describes the relationship that exists between a subnetwork and the smaller subnetwork (or subclasses) instances that are part of its decomposition due to partitioning."	This describes the semantic intent of the information relationship.
ROLE composite "Played by an instance of the <subnetwork> information object type or subtype." component "Played by an instance of the <subnetwork> information object type or subtype."	The relationship is defined between a pair of roles. Roles are used instead of objects because the same relationship can exist between different types of objects, provided that they always play the same roles. In this example, each role can only be played by one type of object, but this is not generally the case.
INVARIANT	A set of invariants is defined (as for the object definitions) in order to put some semantic constraints on the relationship.
inv_cardinalityRoleComponent "At least one instance of the role component must participate in the relationship." inv_cardinalityRoleComposite "One and only one instance of the role composite must participate in the relationship."	These invariants represent the cardinality of the relationship. They say that a composed subnetwork may be related to one or more component subnetworks, and at least one component subnetwork has to exist. However, a component subnetwork cannot be related to more than one "parent" subnetwork through this relationship.

Table 5.3 *(continued)*

Relationship	Description
inv_signalIdentifcation "In a given relationship instance of snIsPartitionedBySn, the information objects playing the role composite and component must have all the same signalIdentification value."	Corresponds to the subnetwork "composition-constraint" property.
inv_roles "In an instance of the relationship, an instance cannot play both roles: composite and component."	The same subnetwork object instance cannot play both roles in the same relationships (i.e., a subnetwork cannot be composed of itself).

policies of the community. The principal active roles involved in supporting this management service are those of the service caller and provider. The service caller role is that of a client of the topology management server (provider role), which performs the actions needed to realize the service, and can be an actual operator making a request through a user interface or an intermediary management system. In Chapter 6, we will see how this decision strongly affects the services that a network operator can eventually offer. The provider role also supports viewing of the resource properties and relationships that have been agreed to in the service contract with the caller. Another potential active role is that of the handler of reporting actions. The actions and operations needed to realize the topology management service are implemented as topology management server software. We note that this service is restricted to be between a single caller and provider, which simplifies the specification at this stage by avoiding the added complication of needing to address the scheduling of several different callers and providers.

The only valid properties for this community are those that are explicitly attributed to it. These can be accessed by the associated service callers and providers and explicit specification is necessary to assure that conformance may be assessed. In the event of a problem, the provider returns an exception that identifies which community obligation or prohibition has been violated, by either the caller or provider, or provides some indication regarding an execution infrastructure problem. The degree to which the caller can understand infrastructure problems depends upon how much shared knowledge is available regarding the infrastructure environment supporting the community. Consequently, we note that detailed infrastructure information should not be passed in exceptions.

5.4.2 Actions

A set of actions is necessary to support the community goal of enabling the creation and deletion of the community resources. These actions are described in more detail than the community purpose and are elaborated in terms of specific action policies and resources. There are three kinds of community actions: create a resource, delete a resource, and report the creation or deletion of a resource. We describe subnetwork creation and deletion actions in further detail within the material that follows.

5.4.2.1 Action: Create Subnetwork

This action creates a subnetwork inside a layer network domain. As illustrated in Figure 5.4, multiple subnetworks can be created inside a layer network domain.

The permissions given to the caller of the service, which will later give rise to attributes and relationships in the information viewpoint, reflect community requirements. These permissions (i.e., action policies) are:

- The caller may provide a user identifier that it will use to uniquely identify the requested subnetwork when it has been created by the provider of the service.

- The caller may provide a user-defined label for the requested subnetwork; this user-defined label cannot be used as a parameter to identify the subnetwork.

These policies ensure that the user identification and label will be optional, as intended.

The corresponding provider obligations are:

- The provider shall, upon success of the *Create Subnetwork* action, return the resource identifier for the created subnetwork to the caller.
- If the caller provided identifier is not unique in the provider context, then the provider shall reject the action.

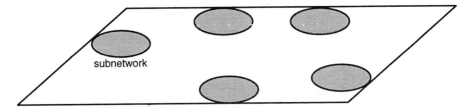

Figure 5.4 Creation of subnetworks.

• If the caller supplied a user identifier, then the provider shall use that identifier as the subnetwork resource identifier when communicating with the caller.

This policy obligates the provider to handle the user identification, as intended. In addition, based upon its policy, the provider may return a list of TTPs that are associated with the created subnetwork. The formalized structure of this action definition, as provided in Table 5.4, is used in the information specification discussed in Section 5.6.

Table 5.4
Formalized Structure of Action Definition—Topology Management, "Create Subnetwork"

topman "Create subnetwork"
"This action creates a subnetwork inside a layer network domain specified by the caller. The caller has the ability to provide a unique user identifier to identify the subnetwork that has been created. The caller has the ability to provide a user label. Depending on its internal policy, its knowledge, and the contract instance, the provider has the ability to return a list of TTPs already associated to the newly created subnetwork. Multiple subnetworks can be created inside a layer network domain."

ACTION_POLICY
PERMISSION inputUserIdentifier
"The caller may provide a user identifier that the caller will use to uniquely identify the requested subnetwork when communicating with the provider."

OBLIGATION successReturnUserId
"The provider shall upon success of this action return a resource identifier for the created subnetwork."

OBLIGATION providerUserId
"If PERMISSION inputUserIdentifier is part of the contracted service, then the provider shall use the user identifier as the subnetwork resource identifier when communicating with the caller."

OBLIGATION rejectUserIdNotUnique
"If PERMISSION inputUserIdentifier is part of the contracted service and if the user identifier is not unique in the provider context, then the provider shall reject the action."

PERMISSION inputUserLabel
"The caller may provide a user label for the requested subnetwork. This user label must not be used by the caller to identify the subnetwork in an operation."

PERMISSION successReturnTTPs
"Based on its policy, the provider may return a list of TTPs that are associated with the created subnetwork (in the case that network equipment has already been installed and is able to autonomously notify its points configuration)."

The ability of this action to return a set of TTPs can give rise to confusion. As we mentioned earlier, the topology management service is generic and can thus be applied to any subnetwork. In the limit it can be applied to an equipment matrix, which is a subnetwork contained within a physical equipment. In this, and only this, case the TTPs are often created autonomously by the equipment, resulting in a subnetwork already bound to TTPs and thus able to support TTP queries. In all other cases, the subnetworks are subsequently bound to termination points using trail and link management operations as briefly discussed in Section 5.3. In the computational viewpoint, this action will be mapped onto an interface and in implementations that interface will be provided by some object. The confusion arises because without knowing some details about that object it is not clear how the interface behavior can be provided. While this may be seen as a problem, there is no real need to understand how a given instance of the interface can be in a position to return TTPs; it is only necessary to accept that in some cases it does. In this book, we describe the service as represented in standards documentation rather than simplifying this particular point in an attempt to avoid potential confusion. We will return to this point later on as we discuss the computational viewpoint.

5.4.2.2 Action: Delete Subnetwork

This action deletes a subnetwork inside a layer network domain at the request of a caller who provides the identifier of the subnetwork to be deleted. This action can only be taken if the particular subnetwork is not associated with any other community resource (i.e., any subnetwork connections, network termination points, links or link connections, or link ends). Thus, no resource other than the subnetwork is deleted by this action. This action will fail if one of the following conditions prevails:

- The subnetwork to be deleted does not exist within the layer network domain;
- One or more TTPs or CTPs are still associated with the subnetwork specified;
- One or more links or link connections are still associated with the subnetwork specified;
- One or more link ends are still associated with the subnetwork specified;
- One or more subnetwork connections are contained within the subnetwork specified.

When the action is successful, the provider shall indicate this to the caller.
For completeness, the remainder of actions required to support the topology management service is summarized in Table 5.5.

Table 5.5
Remainder of Actions to Support the Topology Management Service

Action	Description	Illustration
Create link	This action creates a link between either two subnetworks, or two access groups, or one access group and one subnetwork.	Creation of links (Figure 5.5)
Delete link	This action deletes a link inside a layer network domain. The link shall not contain any link connections. No other resource is deleted by this action.	
Create link end	This action creates a link end (at the edge of a network) bound to a subnetwork specified by the caller. The caller may provide a unique user-defined identifier to identify the link end to be created. The caller may also provide a user-defined label.	Creation of a link end (Figure 5.6)
Delete link end	This action deletes a link end inside a layer network domain. The link end shall not contain any connection termination points. No other resource is deleted by this action.	
Create access group	This action creates an access group inside a layer network domain. The caller may provide a unique user-defined identifier to identify the access group that is to be created. The caller may also provide a user-defined label; the label need not be unique.	Creation of access groups in the layer network (Figure 5.7)
Delete access group	This action deletes an access group inside a layer network domain. The access group cannot be deleted if there are any TTPs associated with it. No other resource is deleted by this action. The caller shall provide the identifier of the access group to be deleted.	
Report resource creation	This action is used by the provider to report on the creation of an instance of a given resource inside a layer network domain. The notification receiver shall be informed by the provider of the identifier of the resource that has been created.	N/A

Table 5.5 *(continued)*

Action	Description	Illustration
Report resource deletion	This action is used by the provider to report on the deletion of an instance of a resource inside a layer network domain. The notification receiver shall be informed by the provider of the identifier of the resource that has been deleted.	N/A

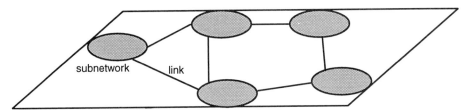

Figure 5.5 Creation of links.

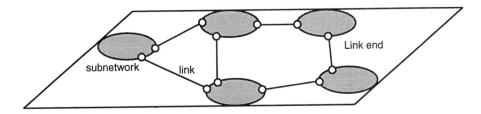

Figure 5.6 Creation of link ends.

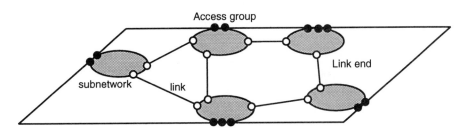

Figure 5.7 Creation of access groups.

5.5 Topology Management—Information Viewpoint

Once the enterprise viewpoint for the topology management service has been defined, we can generate an information viewpoint specification using the concepts introduced in Chapter 4. A complete information specification would provide the objects, attributes, and relationships of interest for the realization of the service. In this section, we will provide a detailed description of two of the information object classes arising from the enterprise specification, the topology management *layer network* and *subnetwork* information object classes. In addition, we will illustrate the full set of information objects and relationships and information object inheritance hierarchy that supports the topology management service.

We will begin with a description of how we may derive the topology management subnetwork information object class. Recalling that management resources are mapped onto information object classes, we see that the subnetwork management resource maps directly to the subnetwork information object. From the preceding section, we saw that the *CreateSubnetwork* action has defined caller permissions and provider obligations related to *inputUserLabels* and *inputUserIdentifiers.* Recalling again that enterprise policies (i.e., permissions and obligations) map onto information attributes and relationships, we see that these caller permissions and obligations should map onto attributes of the topology management subnetwork information object. However, since the subnetwork information object itself does not provide such attributes, we need to subclass the generic subnetwork in order to support the additional service-related information. Additionally, since the generic subnetwork information object itself only specifies potential relationships, we need to provide an explicit relationships clause to specify the actual relationships in which the topology management subnetwork information object can take part.

Thus, as we have seen, we need to create subclasses for the specific topology management service from the generic management resource information objects and relationships. The naming convention used is that the super-class name is prefixed by the name of the service, which is in this case *topman* (an acronym for topology management). The description of the information object class for the subnetwork is further detailed in Table 5.6, and that for layer network domain is provided in Table 5.7.

The complete set of information objects and relationships for the topology management service are illustrated in Figure 5.8. In Chapter 7 we will illustrate how UML can be used to capture these relationships. In existing standards specifications, such figures are typically presented before the specifications; we have reversed the order in this discussion to assist in understanding.

The inheritance hierarchy is provided in Figure 5.9.

Table 5.6
Topology Management Subnetwork Information Object Class

Topology Management Information Object Class	Descriptions
topmanSubnetwork <This information concept is related to the enterprise role subnetwork.>	The name of the object is the concatenation of the service label and its super-type. A reference to the enterprise concern that justifies the definition of this object class.
DEFINITION "This object class is derived from <*subnetwork*>. A subnetwork represents a topological component used to effect routing of a specific signal identification."	An informal definition of the object semantics. Here it just refers to the fact that the object inherits from the subnetwork object, which means that it represents the same concept and has the same definition.
ATTRIBUTE	This clause allows us to define new attributes for the object, which are added to those inherited from the super-class (these latter are not repeated here). These attributes are specific to the defined service and correspond to an enterprise concern defined earlier as a policy rule.
<userLabel>	Here the rule was: 'PERMISSION USER_LABEL', "The caller may also provide a user-defined label..."
RELATIONSHIP <subnetworkIsDelimitedBy> <subnetworkHasSubnetworkConnections> <layerNetworkDomainIsMadeOf>	This clause allows us to be precise about the relationships in which this object can take part.

5.6 Topology Management—Computational Viewpoint

In this section, we will provide one example of a computational viewpoint operation, describe its relationship to information viewpoint concepts, and explain how to read it. To illustrate the creation of an operation from the enterprise and information specifications, we will describe the operation *create-Subnetwork* in more detail. This operation belongs to an interface, *topology-*

Table 5.7
Topology Management Layer Network Domain Information Object Class

Topology Management Information Object Class	Description
topmanLayerNetworkDomain	The name of the object is the concatenation of the service label and its super-type.
<This information concept is related to the enterprise role layer network domain.>	A reference to the enterprise concern that justifies the definition of this object.
DEFINITION "This object class is derived from <layerNetworkDomain>. A layer network domain object represents a transport administrative domain in which all resources pertain to the same G.805 layer."	An informal definition of the object semantics. Here it just refers to the fact that the object inherits from the layerNetworkDomain object, which means that it represents the same concept and has the same definition.
RELATIONSHIP <layerNetworkDomainIsMadeOf>	This clause allows us to be precise about the relationships in which this object can take part. There are two kinds of such relationships: • One inherited from the layerNetworkDomain object where only potential relationships were listed (here only one of the two is required); • One that adds a new relationship that is specific to this management service.

ManagementIfce, which provides all the operations necessary to configure a network topology (comprised of subnetworks, links, and their associated link ends and access groups).

As discussed in Chapter 4, the computational specifications provided in transport management standards (e.g., G.85x) only define computational interfaces and not computational objects. The rationale is to avoid imposing constraints upon future users of such specifications by having to conform to an already defined distribution architecture. Thus, the *topologyManagementIfce,* for example, can be attached to several computational objects, according to a range of design choices. On the one extreme, we could define one computational object per information object corresponding to a management resource. In

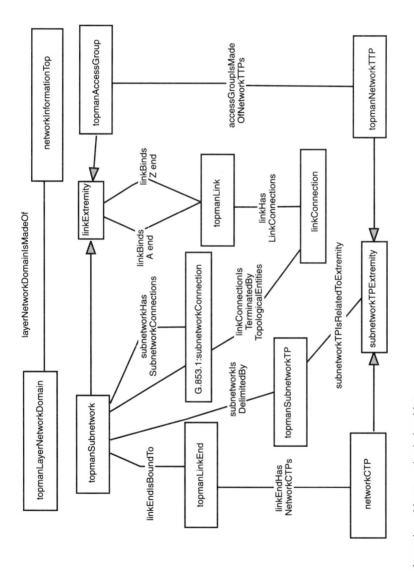

Figure 5.8 Information objects and relationships.

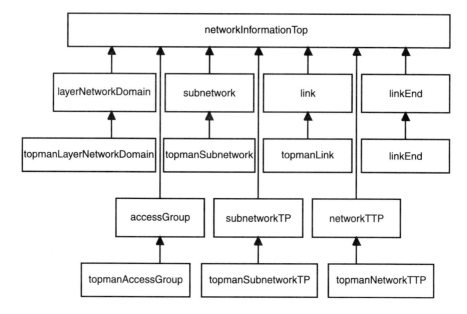

Figure 5.9 Information object inheritance hierarchy.

this case, for a typical network, we could have a set of computational objects encompassing subnetworks, links, and link ends, for example. Thus, the *topologyManagementIfce* we define could be attached to each computational object class, with each one being responsible for its own creation or deletion. At the other extreme, we could define a "Big Administrator" computational object accessible via the *topologyManagementIfce,* whose aim would be to administer the existence of all information regarding management resource objects.

5.6.1 Linkages With Other Viewpoints

Computational interface specifications encompass input and output parameters, exceptions, and behavior (expressed in terms of parameter-matching, pre- and postconditions, and associated exceptions). In this section, we will explain the relationship between the computational specifications and the enterprise and information viewpoint specifications.

In general, as discussed earlier, an action identified in the enterprise viewpoint specification leads directly to the definition of a single operation in the computational viewpoint. Thus, the *createSubnetwork* operation is the computational perspective of the enterprise action *Createsubnetwork.* However, a computational operation may be defined without any enterprise action when it is supporting a computational viewpoint design decision. For example, a design decision has to be made as to whether visibility into management resources will be given or not, which is related to an enterprise obligation. Further elaborating on this point, we consider the enterprise obligation *viewing-Capabilities,* which states that "The provider shall support such viewing of the resource properties and relationships that have been identified and allowed in the service contract with the caller." This obligation leads to a set of query interfaces, one for each resource with a query operation for each type of needed information. These query interfaces will provide access to information concerning management resource identities and properties within the enterprise community supporting the desired service (which we have seen map to information object attributes and relationships), but they do not modify any state. An example involving the topology management layer network domain query interface is provided in Table 5.8.

Computational input and output parameters as well as exceptions make use of information objects and attributes, which have been derived from enterprise viewpoint policy specifications; in particular, the action policy construct where obligations, permissions, and prohibitions associated with either the caller or the provider are expressed. As an illustration, the input parameter *suppliedUserIdentifier* can be deduced from the enterprise requirement captured in the permission labeled *inputUserIdentifier,* "The caller may provide a user identifier that the caller will use to uniquely identify the requested subnetwork when communicating with the provider." The fact that the action policy is a permission can be treated in two ways, dependent upon whether the enterprise contract is discussed point-by-point before its realization and the provider decides up-front whether or not to exercise this permission. If the provider does decide to exercise the permission, it becomes an obligation that is agreed between the client and provider, and the corresponding computational operation needs to contain the associated input parameter. If the enterprise contract remains unchanged because the provider has not made a decision, since it was expressed as a permission, the computational input parameter is defined as optional in the engineering viewpoint.

Table 5.8
Operational Query Interface Relationship With Information Object

Interface Name	Information Object	Attributes and Relationships
TopmanLayerNetworkDomain QueryIfce	<topmanLayerNet workDomain>	<signalIdentification> <resourceId> <layerNetworkDomainIsMade Of, ROLE: element>

Similarly, the output parameter *networkTTPs* corresponds to the enterprise policy labeled *successReturnTTPs*, which is defined as: "Based on its policy, the provider may return a list of TTPs that are associated with the created subnetwork." Again, this policy being a permission, the computational output parameter should be defined as optional. We again note the confusion caused by not being able to see exactly how this interface can act in this way. In hindsight, it would probably have been better to define a separate action to be used only when creating matrix subnetworks (which are implementation objects more obviously associated with an equipment).

The computational exceptions are less trivial to identify. Generally, an exception is associated with each policy violation. For each obligation and prohibition, there is a success case (the policy is respected) or a failure case (the policy is violated).

Once the signature of the operation (the set of input and output parameters and any exceptions) has been clearly defined, its behavior needs to be defined. For this we need a language capable of expressing action; an IDL alone is insufficient for this purpose. The behavior attached to an operation is composed of four parts: parameter-matching, preconditions, postconditions, and exceptions. Whereas the operation signature can be deduced from the enterprise specification, the operation's behavior is related to the information specification. Pre- and postconditions are references to static schemas that have previously been defined in the information specification, stating that the operation has among its objectives that the system (of information objects) for which it is responsible must pass from one state (identified as the operation precondition) to another state (identified as the operation postcondition). Sometimes, for reasons of simplicity and speed, pre- and postconditions are defined "in-line," that is, directly in the operation's behavior instead of in information viewpoint schemas as it should be. An exception is attached to each pre- and postcondition that must be raised in case the condition is not met.

Returning to design decisions, some design decisions have to be made at the computational viewpoint level, which explains why all computational viewpoint constructs cannot be directly derived from enterprise and information specifications. If such a direct derivation were possible, we could automatically generate the computational specification from the other viewpoint specifications. For example, as we saw earlier, a design decision has to be made as to whether visibility to management resources is provided or not. If visibility is to be provided, the relevant query interfaces have to be defined that provide access to information object features (i.e., attributes and relationships). At this stage, the granularity of interfaces is again crucial; a designer could define a single query interface to provide access to the whole set of information objects to be rendered visible through the interface, whereas another equally valid design decision could be to define as many query interfaces as information object types that are made visible. The first alternative offers a simple way to encapsulate legacy systems, whereas the second derives maximum benefit from new distributed architectures such as CORBA. We note that, depending on the complexity of the operation and the communication services, the mapping of the computational operation onto engineering constructs might be a one-to-one (e.g., if CMISE/GDMO is the underlying communication infrastructure) or a one-to-many relationship (e.g., in the case of SNMP where an operation should be decomposed into a number of atomic *gets* and *sets*).

5.6.2 Topology Management Interface Example

As mentioned earlier, we will define one operation in detail for the topology management service, the *createSubnetwork* operation of the operational interface. While this example will utilize the specific computational language constructs and syntax defined in G.851-01 (Appendix 4C), the intent is to enable the reader to understand how such specifications might be written versus learning the specific syntax in detail, since syntax can change without affecting the key underlying concepts. In fact, as we will see in Chapter 7, UML [4] offers a good approach for representing viewpoint specifications and facilitates their usage. In the following subsection, we will introduce the G.851 syntax in more detail to enable an understanding of the computational specification that follows.

5.6.2.1 G.851 Syntax Conventions

The syntax of the operational parameters has been defined through the usage of ASN.1 notation [5] and is called a *supporting ASN.1 production* (an example of which is provided in Table 5.9). In this specification, when an interface is used in an ASN.1 production, the same label will be used starting with a capital letter. The complete ASN.1 production for a query interface can be developed as part of the engineering viewpoint for the target technology.

Table 5.9
Supporting ASN.1 Production

Supporting ASN.1 Productions: SubnetworkId ::= CHOICE { 　　　　subnetworkQueryIfce SubnetworkQueryIfce, 　　　　userIdentifier UserIdentifier};	This means that the SubnetworkId parameter can be represented as either an interface reference (SubnetworkQueryIfce) or a string (UserIdentifier).

Table 5.10
Label Referencing

Full Label Reference	Local Label Reference
<"Rec. G.854.3," INTERFACE: commonReportResourceIfce>	commonReportResourceIfce
<"Rec. G.854.3," INTERFACE: commonResourceIfce>	commonResourceIfce

A further convention relates to label references, which allow us to import information from other sources. Since a full reference may be quite long and hence unwieldy, the definition of a shorter name can be very convenient. We may similarly import interface labels from other documents and assign them an abbreviated name, an example of which is provided in Table 5.10.

5.6.2.2 Topology Management Computational Interface Specification

As mentioned earlier, we will define only one topology management service interface and provide a list of all operations relevant to this interface. In order to increase the readability of the behavior in the operations, we use the following conventions (as is done in existing management services standards documents): parameters are written in **bold** and elements defined in the information specification are written in *italics*. As a representative example, we will provide a detailed explanation of the specification for the *createSubnetwork* operation in Table 5.11.

Table 5.11
Computational Specification for the Create Subnetwork Operation

Operational Interface	Description
COMPUTATIONAL_INTERFACE topologyManagementfce OPERATION { Create Subnetwork Delete Subnetwork Change Resource User Identifier Create Link Delete Link Create Link End Delete Link End	Definition of the computational interface: its name and the set of all operations being described in the specification.

Table 5.11 *(continued)*

Operational Interface	Description

Create Access Group
Delete Access Group
Report Resource Creation
Report Resource Deletion
}
OPERATION createSubnetwork {	

	This operation meets the need of the enterprise action Create Subnetwork.
INPUT_PARAMETERS	Definition of the list of input parameters of the operation.
layerND : LayerNetworkDomain ::= (topmanLayerNetworkDomainQueryIfce);	LayerND is the name of the parameter. LayerNetworkDomain is the type of the parameter that is a interface reference to (topmanLayerNetworkDomainQueryIfce).
suppliedUserIdentifier : UserIdentifier; -- zero length string or 0 implies none -- supplied	The comment in this definition means that this parameter is optional. This corresponds to an enterprise permission allowing the caller to provide a user identifier (or not).
suppliedUserLabel : GraphicString; -- zero length string implies none -- supplied	
OUTPUT_PARAMETERS	Definition of the list of output parameters of the operation.
subnetwork : SubnetworkId;	The result of the operation will be to create a subnetwork object. This object is of SubnetworkId type, which has been seen in the ASN.1 definition as either an interface reference or an identifier string.
networkTTPs : networkTTPIds ::= SET OF TopmanNetworkTTPQueryIfce; -- the parameter networkTTPs is only -- valid when <COMMUNITY: topman, -- ACTION: create subnetwork, -- PERMISSION: successReturnTTPs> -- is used	
RAISED_EXCEPTIONS	List of all exceptions of concern to this operation.
userIdentifierNotUnique : NULL; failureToSetUserIdentifier: NULL; failureToCreateSubnetwork: NULL; failureToCreateNTTP: NULL; failureToAssociateNTTP: NULL;	The exceptions are just listed here with a name and a type; here the NULL type means that the choice is left to the implementers.
BEHAVIOR	Definition of the operation behavior.

Table 5.11 *(continued)*

Operational Interface	Description
SEMI_FORMAL PARAMETER_MATCHING	The first part of the behavior description is the parameter-matching clause. This is where the "glue" between the information specification and the computational specification is formally represented.
layerND : <INFORMATION OBJECT: topmanLayerNetworkDomain>;	layerND is the name of the input parameter. This parameter is bound here to the information object type topmanLayerNetworkDomain. This allows us to add to the parameter syntactic definition a strong semantic that can be found in the information specification.
suppliedUserIdentifier : <INFORMATION ATTRIBUTE: resourceId>; suppliedUserLabel : <INFORMATION ATTRIBUTE: userLabel>; subnetwork : <INFORMATION OBJECT: topmanSubNetwork >; networkTTPs ELEMENTS : <INFORMATION OBJECT: topmanNetworkTTP>;	Such a parameter refers not to an object type but to an attribute type.
PRE_CONDITIONS	The second part of the behavior description is the definition of the pre- and postconditions. This is also a reference to the information specification. It allows us to specify the conditions that have to be satisfied before the operation can be executed (preconditions) and the conditions that will be true after the operation has been executed.
inv_uniqueUserIdentifier "The **suppliedUserIdentifier** value shall not be equal to the *resourceId* value of any *element* in the *<layerNetworkDomainIsMadeOf>* relationship where the **layerND** refers to *container*."	Each pre- (or post-) condition has a name that is an invariant name. For the ease of readability of specifications these invariants are defined in the computational specification. The other way to do it (as explained in Chapter 4) is to refer to information static schemas, which correspond to this invariant. The computational parameter is in **bold** and the information concepts (attributes, roles, objects, or relationships) are in *italics*.

Table 5.11 *(continued)*

Operational Interface	Description
POST_CONDITIONS inv_agreedUserIdentifier "The *resourceId* value of the *topmanSubnetwork* referred to by **subnetwork** is equal to the **suppliedUserIdentifier** value, if it is supplied."	This can be read as: The attribute 'resourceId' of the 'subnetwork' object that is being created will take the value of the **suppliedUserIdentifier** passed as that parameter, if there was one.
inv_existingSubnetwork "The **subnetwork** and **layerND** must respectively refer to the *element* and *container* of the same *<layerNetworkDomainIsMadeOf>* relationship." -- Note: the following -- invariants are only needed if the -- networkTTPs appear automatically -- when the subnetwork is created	
inv_existingNetworkTTP "The **networkTTP** and **layerND** must respectively refer to the *element* and *container* in the same *<layerNetworkDomainIsMadeOf>* relationship."	
inv_nttpAssociated "The **networkTTP** refers to *extremity* of the *<subnetworkTPIsRelatedToExtremity>* relationship where the *abstraction* (which is subnetworkTP) is also the *element* of a *<subnetworkIsDelimitedBy>* relationship where the **subnetwork** refers to the *container*."	
EXCEPTIONS	The third part of the behavior description is the exact definition of the exceptions. It defines in which case an exception will occur. Each exception corresponds to the fact that a pre- or postcondition has been violated and then can give a semantic reason for this case of failure of the operation.
IF PRE_CONDITION inv_uniqueUserIdentifier NOT_VERIFIED RAISE_EXCEPTION userIdentifierNotUnique;	This can be read: If the caller gives a user identifier that is not unique the operation will fail; it corresponds to an enterprise obligation (see next paragraph).

Table 5.11 *(continued)*

Operational Interface	Description
IF POST_CONDITION inv_agreedUserIdentifier NOT_VERIFIED RAISE_EXCEPTION failureToSetUserIdentifier; IF POST_CONDITION inv_existingSubnetwork NOT_VERIFIED RAISE_EXCEPTION failureToCreateSubnetwork; IF POST_CONDITION inv_existingNetworkTTP NOT_VERIFIED RAISE_EXCEPTION failureToCreateNTTP; IF POST_CONDITION inv_nttpAssociated NOT_VERIFIED RAISE_EXCEPTION failureToAssociateNTTP;;}	Note that the exception is actually specified in the raised exceptions section.

5.7 Summary

In this chapter, we applied the generic principles introduced in Chapter 4 to illustrate the construction of a specification for a simple management service. In doing so, we attempted to stress the relationships between the different viewpoints and the design decisions taking place in each viewpoint. We discussed the apparent complexity that arises from the explicit nature of the specification technique and why incomplete specifications with many implicit details are not simple specifications. We also noted the confusion that can arise when too much emphasis is placed on how an implementation might produce the interface behavior specified and that this is the difference between creating a specification and understanding how an implementation might work. In Chapter 7 that follows, we will once again make use of this simple example to demonstrate how we may express viewpoint specifications in UML [4].

References

[1]   ITU-T Rec. G.852.2 "Enterprise Viewpoint Description of the Transport Network Resource Model," Mar. 1999.

[2]   ITU-T Rec. G.853.1 "Common Elements of the Information Viewpoint for the Management of a Transport Network," Nov. 1996.

[3] "A Network Management Forum Business Process Model," Network Management Forum, GB 908, Nov. 1997.

[4] Lee, R. C., and W. M. Tepfenhart "UML and C++, A Practical Guide to Object-Oriented Development," Englewood Cliffs, NJ: Prentice-Hall, 1997.

[5] ITU-T Rec. X.208 "Specification of Abstract Syntax Notation One (ASN.1)," 1988.

6

An Overall Approach to Modeling

6.1 Introduction

In Chapter 1, we provided some motivation for the general principle of establishing a unified model for describing all the entities that play a role in a network. Here, we will elaborate on these motivations further, after first discussing more completely what we mean by the notion of a unified model (Figure 6.1). When we consider what such a model would look like, it becomes clear that we are not referring to a single massive information base that captures all information relevant to every aspect of telecommunications. Nor are we necessarily referring to a common notation that is equally applicable to the specification of all aspects of transport, management, and control. In fact, what we are actually suggesting is characterization of the relationships between existing and developing models in each of these domains. Throughout the remainder of this chapter, we explore the relationships between transport and management, motivate the need for further unification, and outline an ODP-based approach for establishing a unified modeling framework. Full integration with the control domain is left for future work, though we will provide some rationale as to why we believe that the concepts described herein are equally applicable to that problem domain.

6.2 Diversity of Modeling Approaches and Implications

As discussed in the preceding chapters, there are a variety of modeling techniques being used to assist in the design of transport networks and their associated control by either management operations or signaling systems. In addition,

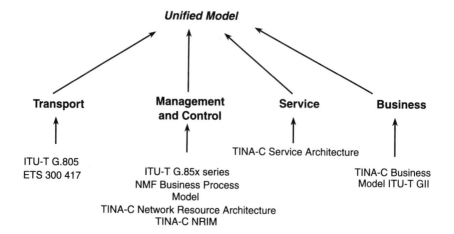

Figure 6.1 Domains of interest for a unified model.

transport protocols and interfaces have been developed and associated with one or more of these models. Within the transport domain, we use the transport functional modeling approach described in Chapter 3, which is based upon entities and relationships described within ITU-T Rec. G.805 [1] and the ETS 300 417 series of standards [2–8]. Within the network management domain, we use the augmented RM-ODP-based approach described in Chapter 4, based upon the RM-ODP framework of abstractions [9] and the ITU-T Rec. G.85x series [10–13] standards. Within the control domain, the modeling of control (i.e., signaling) aspects for intelligent networking is based upon RM-ODP concepts within evolving TINA-C specifications [14]. The existence of such diversity has caused considerable ambiguity at the boundary of domains where notation and concepts utilized differ. For example, for a given equipment, the relationship between a GDMO object model for management and its transport functional description is not obvious (Figure 6.2). In particular, there are differences not only in the notation that is used, but also in the way that associated behaviors are expressed and modeled.

Note that we have adopted a drawing shorthand for depicting a transport interface with embedded management communications in Figure 6.2 (as shown in the inset, the rectangular symbol represents these flows being combined according to the rules of functional modeling). This correspondence is depicted in the inset. In the following sections, we will provide some examples of key network maintenance parameters, availability and QoS, that are affected by these multiple domains of interest.

6.2.1 Availability

Overall network availability is strongly affected by the availability of the management communications, especially when management and transport use shared

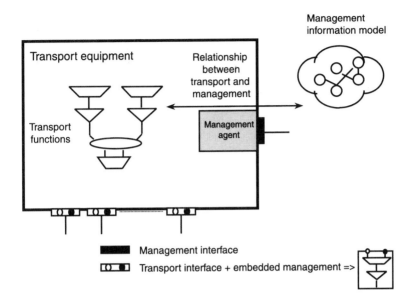

Figure 6.2 Notational differences between equipment functions and associated management model.

communications mechanisms, including those embedded within equipment. When management interactions use transport networks for their communications needs, there is a relationship between the managed network and the network that conveys management information. Currently, when we calculate the service availability, we tend to assume that the management communications network is always available, that is, safe for the interaction. This ideal case is usually considered when a programmer designs an application that will run on a given centralized system. It is also assumed to be the case for interactions between systems using a good transport protocol (OSI layer 4, or equivalent), although that ideal case for intersystem communications is sometimes difficult to guarantee in practical networks. In the case of transport management, the problem arises from the fact that the purpose of the interaction is to control the transport itself. The main issue is that the targeted 100% availability may not be achieved, especially when the management interactions themselves contribute to unavailability. There are several classical examples of scenarios in which this is the case. One of the most famous corresponds to simultaneous transport and management function failures, due to shared hardware or software, leading to an inability to communicate failure status. Another classical problem is the case where management interfaces share communications resources with managed transport networks, leading to the same situation as internal system failure.

Traditional modeling approaches characterize transport and management aspects separately. Since these approaches model the relevant availability characteristics on a piecemeal rather than holistic basis, potentially important availability issues arising from the interdependencies of the transport and management characteristics may remain hidden. In other words, modeling management and transport independently of each other carries with it the implicit but often inaccurate assumption that the abstracted domain (transport with respect to management, management with respect to transport) is completely reliable.

6.2.2 Quality of Service

In supporting QoS objectives for various services, it is necessary to ensure that management information is reliably communicated and relationships between needed resources are clearly specified. Associated QoS parameters may be used to guide the selection of optimal transport protocols, communications resources, management of communications resources, and sequencing from the point of transport signal service request to transport signal service receipt. Such parameters also include temporal constraints (e.g., connection establishment delay, transfer delay), which need to be communicated to all the involved resources and activities along the communications path. Algorithms, which transform QoS parameters into a set of activities that act upon resources to guarantee that QoS objectives are met, can be quite complex, especially due to the overlap of transport and transport management activities. In fact, the problem is so complex that current circuit networks are invariably somewhat overengineered while best effort networks often tend to underutilize their resources.

6.3 Rationale for a Unified Modeling Framework

In order to establish a unified modeling framework, it is necessary to capture the set of relationships crossing the boundaries of existing models. Examples of the kinds of relationships between models that need to be precisely captured include those between equipment functional models and their representation for equipment management (in terms of information viewpoint specifications); between equipment-level and network-level management models; among models of network management services and network- and equipment-level management models; and between management and control models. Primary drivers for establishing these relationships encompass the blurring of traditional transport and management domain boundaries, network operators' need to understand overall network behavior, and the continued industry drive toward interoperability among different vendors' equipment.

The current separation of transport and management behavior has been largely driven by traditional assumptions regarding the respective roles of these domains. It has been traditionally assumed that transport is concerned with the movement of information but not directly with the information content itself. Conversely, it has been assumed that management is associated with the information that passes between and among transport equipment and management software to control and supervise transport network behavior, but not directly with the transport of client information. However, as networking capabilities have evolved, management and transport functions are becoming increasingly intertwined and their boundaries increasingly blurred. For example, to a large extent, support for equipment supervisory functions is now built directly into SDH/SONET transport signals as management overhead. We also see that, in the interests of improved performance, management functions that reroute traffic around failed equipment and links (e.g., ring protection switching) are increasingly embedded within transport equipment. Thus, specification of equipment behavior necessarily involves an integrated treatment of management behavior.

Similarly, specification of network behavior necessarily involves an integrated treatment of management behavior. From a network operator perspective, or for that matter, from the perspective of any client of the network's functionality, it is essential to understand the behavior of the network as a whole (which includes both management and transport characteristics). This is especially true for cases in which some portion of the supervisory capability is extended to the client, as in end-customer control scenarios (e.g., virtual private networks). This case, in particular, also illustrates the need to capture the relationships that exist between the users, or clients, of network services and the providers, or servers, of the services. In other words, there is a need to formally characterize the effects of events in a client or server layer on the corresponding server or client layer.

Finally, as we discussed in Chapter 2, the goal of interoperability among various vendors' equipment and between different operators' network domains cannot be achieved without addressing the relationship between network- and equipment-level transport and management functionality, regardless of how it has been distributed within networks and equipment.

Referring back to our discussion of overall availability in Section 6.2.1, we must be able to model both transport and management within a common framework to properly understand the key relationships and dependencies that could lead to availability impacting deadlock scenarios. Only then can we anticipate the service availability impact on the whole network of any given interaction that could take place in a management application and its communications facilities. Similarly, a unified modeling approach would facilitate realization of target QoS objectives for services by relating the desired QoS to the performance of the associated network resources.

As stated earlier, the notion of a unified model does not imply a monolithic semantic expression of all potentially related characteristics but a normalization of those relationships that may be important for a particular endeavor. In particular, many modeling-related activities will continue to characterize aspects unique to a particular domain. An important example of this is the transport functional modeling approach, described in Chapter 3. Because this approach is so specifically dedicated to the expression of transport-related functions, it has developed domain-specific semantics that are rich and efficient for expressing transport network architecture specifications.

Similarly, to establish a general approach that is capable of describing any interesting relationships among network functions, it follows that semantics rich and powerful enough to handle the implications of this responsibility are also required. This implies that we need to draw upon the generic body of work throughout the computer and telecommunications industries for descriptive techniques for system functionality. Further, for cases in which domain-specific semantics are appropriate, we need to understand and formalize the relationships and implications with respect to other domains. Ultimately, the scope of a unified approach to modeling should include the ability to model any part of the transport domain for not only equipment functional design but also for network planning, network services, and service access design purposes. Such a model should allow us to describe any transport management domain, including both equipment and network management application designs, and be independent of control by management or signaling protocols and of the circuit, or packet, nature of the network.

6.3.1 Modeling Shared Communications

In Section 6.3, we discussed the interdependencies between transport and management functionality. Our first observation is that architecture, modeling concepts, and notation that are used to describe transport, management, and control activities can be used equally, and more or less interchangeably, depending on which aspects are interesting for specification. Thus, when there is a need to describe the communications aspects of a management or control protocol, there is no need to change our approach toward modeling the transport functionality simply because we are interested in the transport of management information.

Management communications are frequently message based and involve some protocol stack, such as the OSI stack. Layers 1 to 4 of the OSI stack can be compared to a transport network layer. Specifically, we can model the communications aspects of this management protocol, using the functional modeling approach of Chapter 3, by introducing trail termination functions representing the OSI transport connection between the management applications. This matches the needs of the OSI transport protocol to the available physical transport layer (i.e., it indicates that the transport network should provide Layer 4 functionality). The implementation of this trail termination function may be quite complex, given that it adapts the requested QoS for the management interaction to the available QoS from the network. However, modeling this functionality is quite simple. The coupling to the management application is via an adaptation function that matches the management information source to the transport protocol (i.e., relates Layer 7 to Layer 4 in the OSI structure). Again, while this functionality is not trivial to implement, it is simple to model. These concepts are illustrated in Figure 6.3.

This management transport adaptation functionality can, of course, be located within the transport equipment itself, as illustrated in Figure 6.4.

This process of recursion can be continued until it can be seen that the elementary transport function itself must include a management interface.

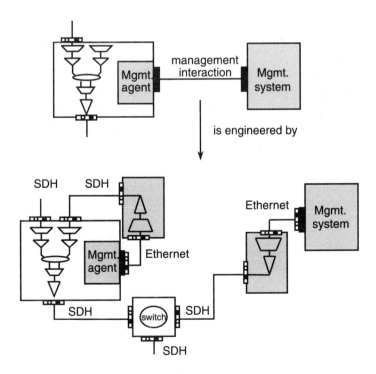

Figure 6.3 Management interactions via managed transport networks as external communications systems.

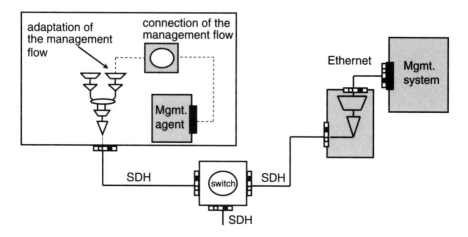

Figure 6.4 Management interactions via managed transport networks as internal and external communications systems.

Thus, we can state that a transport function is itself decomposable into a transport and a management part, with corresponding communications between them. Internal communication between management parts and/or transport parts will often use communication facilities that are shared (e.g., a back plane), and these internal communications can also be easily modeled, as illustrated in Figure 6.5.

Just as we argued that there is no need to introduce new modeling techniques for management of communications, we argue that there is no need to introduce different techniques for modeling internal management activities

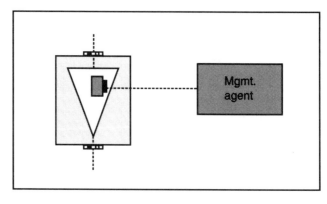

------- internal interaction

Figure 6.5 A trail termination function including some management process.

(Figure 6.6) as opposed to external management activities. We can quite easily use the techniques of Chapter 4 to specify the equipment internal management processes.

We have now arrived at one aspect of a unification of models whereby the interactions between management communications and shared communications can be described together. There is still a strong conceptual separation between the management and transport parts; however, the common communications facilities that create a strong interaction between them may be accurately modeled. This now allows us to make more precise statements about the overall behavior of the network.

6.3.2 Modeling Shared Knowledge

We showed in the previous section how to relate transport network models with associated control and management models. It would appear that the ultimate benefit of this would be to have one and only one way (notation and associated semantics) to describe all involved activities. Does this imply that we must have one and only one model? As we suggested earlier, the answer is clearly no. For example, we cannot expect the client of a bank to care about the bank's internal organizational structure when making a transaction. The bank, likewise, is not interested in the particular means used to convey the information needed during the transaction and may well subcontract the rele-

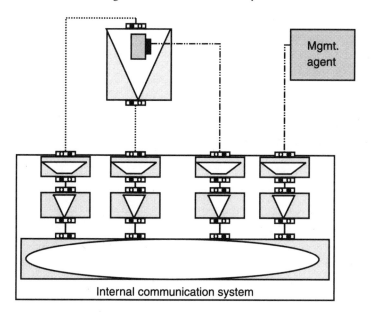

Figure 6.6 Internal equipment communications transport via an internal trail.

vant information services without any interest in understanding how those services are actually provided. This observation leads us to realize that fundamental differences among models occur when they are capturing different aspects of shared elements, and such differences remain regardless of whether these models are using the same modeling language. In this chapter, when we speak of different models, we are referring to the semantic difference described previously, rather than the syntactic difference that would arise from the use of different modeling languages.

Let us once again consider our example of the bank, this time in a transaction that involves the buying or selling of stocks. In this case, the transacted resources are not the property of the bank, and the bank must establish a subcontracting arrangement with a stockbroker in order to complete the transaction. While the broker, bank, and client have a mutual lack of interest in the "private" actions each needs to take in order to perform the transaction, they must nevertheless have a shared notion of stock, for without this there could not be a transaction at all. Therefore, some essential knowledge must be shared between both the client and service provider. In the remainder of this section, we will further explore the concept of shared knowledge, the coupling created by the shared knowledge between various models, and the associated potential for inconsistencies. We will also see how shared knowledge can lead to strong accidental couplings that can overly constrain designs.

As discussed in Chapter 4, a service is considered to be a contract between two roles, those of a client and a provider, regarding an activity to be performed on behalf of the client. The shared knowledge, therefore, represents the subset of resource properties needed for the service requirements to be correctly negotiated between the two parties and for the associated actions to be correctly performed (Figure 6.7).

In order to realize the stock transaction service, the bank may require additional resources to complete the transaction and may also have to make use of additional knowledge it has about the shared resources. However, while the bank needs to know how to deal with brokers, the customer does not require this knowledge. Other resources may also be involved, which are either completely private to the bank, or for which the bank is a client. For example, the bank may have an internal account that it uses to process stock transactions, as well as arrangements with the broker for delivery of the actual stock certificates. A possibility for accidental coupling arises should the bank allow the broker to make transactions directly on this internal account, as changes to that account would appear in the bank's interface to the broker.

In an information system, extra knowledge about shared resources appears as additional properties of these resources. Obviously, to maintain global consistency, the added properties cannot contradict the shared properties. As we

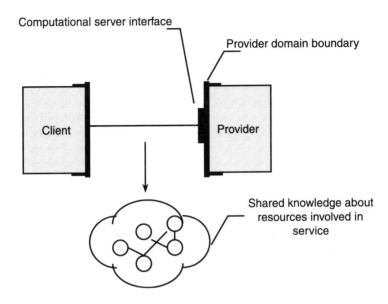

Computational server interface

Provider domain boundary

Client

Provider

Shared knowledge about resources involved in service

Figure 6.7 Shared knowledge between client and provider of a service.

suggested earlier, the bank also assumes a client role in dealing with the broker, and the shared resources here may also include other properties that correspond to their usage; for example, there may be additional states or relationships with other resources. Just as for the client-bank interface, the additional properties that are introduced here cannot contradict the shared properties. These points are illustrated in Figure 6.8.

Up to this point in the discussion, the properties we have been considering have not depended on the context in which they are being used. A stock certificate, after all, does not change its value depending on who owns it. However, some resources have properties that are context-dependent, and in this case, the provider has to add the context to the resource property, and both must be considered for consistency purposes. The most obvious example is a resource that may be shared among several clients and considered as free, reserved, or in use. When a client reserves the resource, the resource must appear to be in use to all other clients. Therefore, the simple resource properties of reserved and in use are now not sufficient and the provider must maintain this extra knowledge (as additional properties) to capture the context dependent use of the resource. The context annotation is an additional property maintained in the provider domain that respects the shared knowledge between the provider and its clients. An example of this could be a bank deposit box, where the distinction between reserved and in use is important between the bank and the customer but is not visible to other customers trying to get a deposit box.

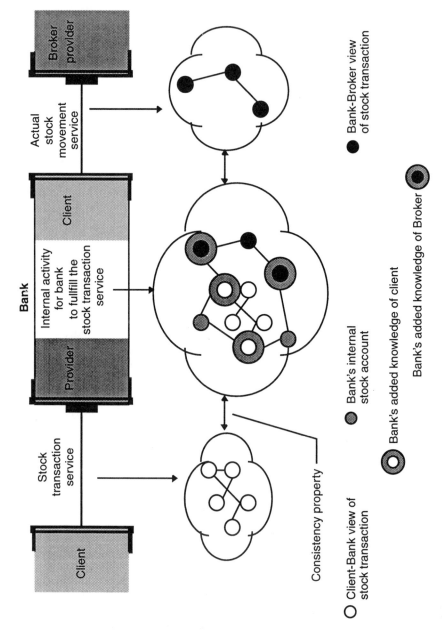

Figure 6.8 Generalization of distributed knowledge.

Notice that while we have described a bank—with an implied physical separation between client, bank, and broker—the difference in shared knowledge is not caused by physical distribution but is actually caused by the abstraction boundaries between the different domains of interest. The discussion would be unchanged if the client, bank, and broker were all located in the same room. The constraints expressing global consistency are invariant and therefore need to be maintained over time. Clearly, physical distribution of resources and activities makes these invariants more difficult to guarantee and an infrastructure that masks distribution effects will reduce the risk of violating consistency rules.

Considering shared resources further, we come to understand that the design process creates resources, and hence designers create those shared resources. This would not be a problem if a single authority designed the complete model and the implementation of the various parts had been delegated. However, this is clearly impossible because the various domains are typically designed in isolation. For example, the equipment that implements telecommunications service access points may be part of the service provider's domain model but the transport equipment inside that domain will not be part of that model, because it would have been designed independently (entirely outside the provider domain). Designers are also free to share more knowledge than absolutely necessary and this leads to accidental couplings that make future changes difficult. Usually such accidental couplings are the results of convenient optimizations. For example, it might be convenient for the management system to know how the vendor names service access points since this saves a name translation. Later on, of course, this optimization could cause a disaster when the vendor wants to change the internal naming scheme for any reason. In the following subsections, we will provide some examples that illustrate these concepts.

6.4 Examples of Shared Communications and Knowledge

In this section, we will provide several scenarios illustrating applications of the concepts of shared communications and shared knowledge. These scenarios will encompass integrated transport and management, link partitioning, and so-called virtual private networks.

6.4.1 Integrated Transport and Management

Let us first consider an example that relates functional transport domain specifications and management information models. Referring to Figure 6.9, we will

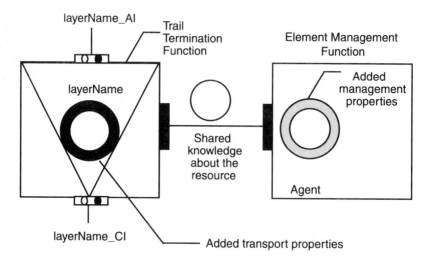

Figure 6.9 Shared knowledge between a trail termination function and its management representation.

consider a trail termination function where we are primarily interested in its management aspects (i.e., the management interface indications and associated shared knowledge). The trail termination function is the client of management services provided by an element management function (the unspecified EMF of Chapter 3), and some of its properties will constitute part of the shared knowledge among them. The trail termination function adds some properties and resources, such as detailed information about the transport interfaces, in order to provide its service. Likewise, the element management function is a client of the management services provided by the trail termination function and these properties also form a part of the shared knowledge among them. The element management function, of course, will add some additional properties and resources of its own. Clearly, only the shared information between these two functions needs to be modeled.

As described in Chapter 2, the TMN logical layered architecture refers to element management, network management, and service and business management. Such an architecture assumes that each layer has its own domain of interest and is basically independent of the organization of the layer below it. Nevertheless, each layer has interfaces to the layer below it, and these interfaces carry the shared knowledge that is necessary and sufficient to enable the layer to accomplish its purpose. We cannot sufficiently stress the importance of "necessary and sufficient." This principle applies until we reach the lowest layer, which is inside equipment where the shared knowledge between the element management function and the trail termination function resides. The trail termination function properties are therefore part of shared properties that are invariant, and the TMN is responsible for ensuring their consistency.

6.4.2 Link Partitioning

The serial partitioning of links offers another scenario involving knowledge shared among several layers, with the attendant problem of ensuring global consistency of the shared properties. As we recall from Chapter 3, a link represents a fixed transfer capacity between two subnetworks that may be provided by trails. From the perspective of the client layer, a link defines an abstract amount of capacity and a quality of transfer service indication, which is derived from the trail quality of transport. In our models, we can provide the link via a technology-independent trail, which we will call an *infrastructure trail*. The infrastructure trail represents the way to transform the client desired QoS into selection of an actual trail that offers the best underlying technology for providing the required QoS. This coupling binds some services very tightly to physical network resources, and subsequent sections will describe ways of reducing this coupling. Chapter 8 will discuss the engineering difficulties of constructing management interfaces between technologies using different management protocols. While the models we employ are always logical, it is very tempting to always associate a technology with the layer. The infrastructure trail concept enables us to delay the selection of the technology. Figure 6.10 shows a portion of an ATM virtual path network, where the ATM links are provided by transport-independent infrastructure trails.

These trails are, in turn, provided by SDH or cell-based physical transport system resources. The SDH trail resources represent shared knowledge with the infrastructure trail manager, which presents the actual resources used by the ATM link. The infrastructure trail is responsible for hiding the shared physical transport system resources from the ATM layer. In this way, we delay binding to a particular technology and enable the ATM layer to take advantage of any physical transport capable of offering the required QoS, which enables us to make more cost-effective decisions regarding the selection of physical transport.

6.4.3 Technology-Independent Networks

From the point of view of the modeling concepts we have introduced, we can conceive of whole networks that are transport technology-independent. In order to do this, we also have to consider subnetworks as being technology-independent. This is in fact the case because the function of a subnetwork is to make associations between ports on its boundary. When those ports belong

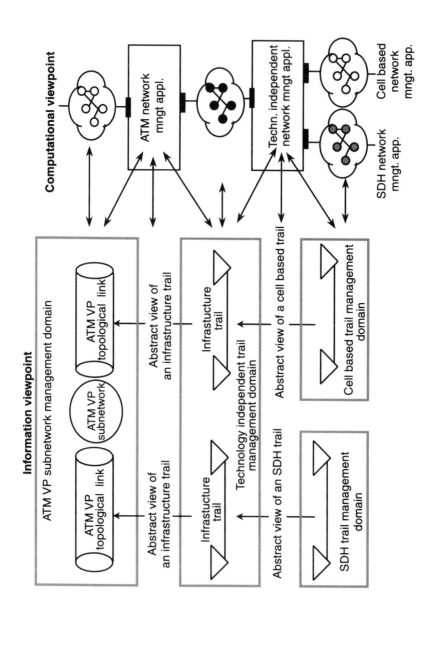

Figure 6.10 ATM virtual path network example of link partitioning to support QoS objectives.

to transport technology-independent links, then the subnetwork itself is also technology-independent. Functional models are logical and thus independent of actual physical implementation. However, a functional model of an SDH network, for example, is not a transport technology-independent model. The technology-independent aspect comes from the infrastructure trail of Figure 6.10. This provides an abstraction of another network, which can itself be of a technology specific or abstract nature (as illustrated in Figure 6.11). Usage, control, and management of the resources of such networks is often provided as a service to an end client, and such an application is often known as a virtual private network (VPN). The network is virtual because what is being managed is the technology-independent layer, not a physical layer, so the topology offered need not mirror a physical underlying network. It is considered to be private, not because of enhanced security or privacy, but because it is under the control of the client, not the supplier. For a given network, the same abstraction process may be used to construct other networks and, obviously, several networks may be constructed from any given one, thus allowing several apparently independent networks to share the same underlying physical resources should the provider desire to do this. Any sharing would be modeled by link partitioning as discussed earlier.

6.5 Relating Transport and Management Modeling Views

Let us return to the unified modeling of transport and management. In Section 6.3, we saw how we can use the same specification techniques to describe the transport of management information and the management of transport

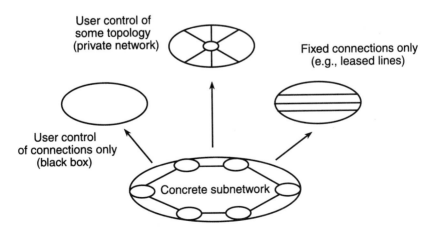

Figure 6.11 Virtual networks.

information. We will now show that both transport and management can be handled using the viewpoint techniques introduced in Chapter 4. We will now consider a specific example that illustrates the correspondence between transport and management models based upon the atomic function fault process application described in Chapter 3. Specifically, we will consider the trail termination sink function (TT_Sk), which is defined by its input and output signals, and internal processes including detection of events and consequent actions toward the client adaptation, the trail termination source function (TT_So), and the management function. When the trail termination sink function (Figure 6.12) detects a failure in a trail, such as an SDH path, a trail signal failure (TSF) indication is sent toward the client's adaptation to inform them about the server trail failure condition.

As discussed in Chapter 3, a management function will receive management events, process them, and any consequent action toward external management systems would appear as messages to the equipment management interface. Using the management information specification techniques described in Chapter 4, the management specification of the behavior that governs failure detection is expressed as an invariant in an information object called *sinkSDHTrailTerminationPoint*, provided in Table 6.1.

Comparing the transport functional modeling approach and the viewpoint approach for modeling management, we can see that the transport domain specification of equipment behavior tends to specify interfaces and is thus computational viewpoint oriented, while the management domain specification of behavior is more information viewpoint oriented. This is not a problem as long as we ensure that the relationship between specifications is consistently and unambiguously expressed. Both the specifications must eventually describe the information objects common to both, which represent the shared information of Figure 6.9.

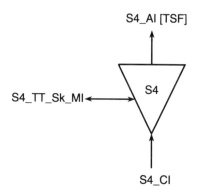

S4_AI [TSF]

S4_TT_Sk_MI

S4

S4_CI

Figure 6.12 VC-4 trail termination sink function—relevant aspects.

Table 6.1
sinkSDHTrailTerminationPoint

sinkSDHTrailTerminationPoint
DEFINITION "This information object class is derived fromTTPSink." INVARIANTS inv_SignalIdentification "The signal identification is to be from the SDH path layers (i.e., VC4, VC3, VC2, VC12, VC11) or section layers (i.e., OS, RS, MS)." inv_sinkPartDisabled "The sink part of the trail termination is disabled if the trail termination function (TT_Sk) sends the consequent action 'TSF' downstream to its associated adaptation function."

In Chapter 4, we discussed the stream interface type and mentioned its use for modeling transport interfaces. We can therefore use computational objects to represent the functional model entities, using stream interfaces to model traffic and operations interfaces to model management interfaces. In subsequent sections, we will bridge the differences between the management and transport domains by considering a transport model atomic function as a computational object, which can be visualized by depicting operations interface symbols directly on the atomic function symbols.

Referring again to our illustrative example, we can now see that a computational object, corresponding to the *sinkSDHTrailTerminationPoint* object, must exist internal to the equipment, and associated management indications need to be sent downstream. The functional model of the equipment must therefore have a notification interface available to the TT_Sk function to bind to the computational management object. The interaction between the TT_Sk function and the management computational object always takes place inside equipment and is considered to be an equipment implementation aspect that is not subject to standardization.

6.6 General Framework for Transport Domain

When we are working in the engineering and technology viewpoints, the decomposition of general transport facilities needs to be considered. So far, we have only considered the characteristic information of a layer network to reflect physical characteristics of the signal (e.g., its bit rate, format, or other such properties).

In previous sections, we discussed how functional modeling concepts can be used to describe technology-independent transport and how these enable us to describe virtual private networks. It should now be a small step to consider characteristic information to be the abstract property of the signal as originally intended rather than properties having a physical reality, such as bit rate and format. As described in Chapter 3, the most important property of characteristic information is that of connection compatibility; that is, two signals can only be connected when they have the same characteristic information. This property is preserved when we consider characteristic information to be defined by signal semantics; in general, this simply means that if the two signals have equivalent semantics, they can be interconnected. For example, we can display the same television program in different countries even though different physical TV standards are being employed. In the same way that abstract trails enable us to model transport technology independence, abstract characteristic information enables us to model the conversion between signals having the same semantic content but different physical representations. In the following figures, we use the term *characteristic information* <N>, by which we mean signals have the same semantics with possibly differing representations. Because of the common assumption that characteristic information is about concrete properties, within TINA-C the term *signal identification* has been used to represent the original notion of abstract characteristic information.

In the general case depicted in Figure 6.13, the architectural intention is to model information transport between usage applications (i.e., any information processing, storage, or transport application) independent of the technology used to realize the transport. At this level of modeling, the only interesting aspects are the interactions that are needed between different usage applications. Because most aspects of transport are irrelevant to the computational viewpoint, we only need to model the usage domain adaptation if multiplexing facilities are used, since it can be important to know when and how transport facilities are being shared. Here we consider a usage domain to be delimited, for example, by the hardware supporting the usage application and communications facilities.

The transport domain becomes more complex in the case of cooperation among transport domains supporting different signal representations, as shown in Figure 6.14. The transport domain that manages the transposition between the characteristic information <4> and <5> includes an interceptor domain, which functionally corresponds to the ODP interceptor engineering object mentioned in Chapter 4. Such a domain is modeled by two adaptation functions connected back-to-back (shown here as adaptations A4 and A5), which adapt the different representations via a common, unspecified representation. (We note that a distributed interceptor requires standardized characteristic information.) One common use of the interceptor is to model bridges and gateways

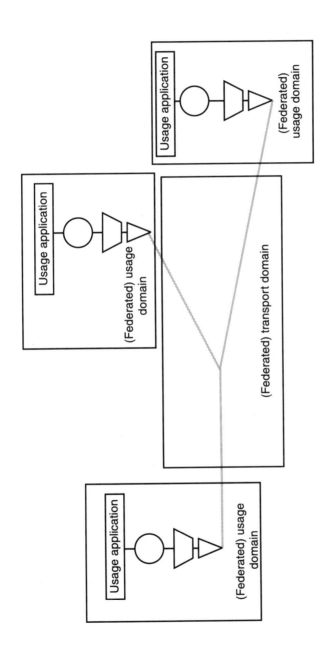

Figure 6.13 General transport facilities.

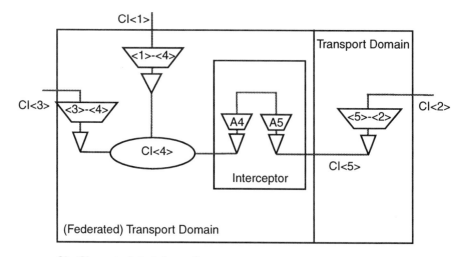

CI - Characteristic Information

Figure 6.14 Cooperation with another transport domain supporting a different CI.

between different protocols. Each protocol corresponds to a different representation, and of course, there are limits to the degree of conversion as well as to the ease of conversion, as will be discussed in Chapter 8.

Because the atomic functions of the transport functional model preserve signal semantics while changing signal representation, we can consider trails to have identical signal semantics with different signal representations at their terminations. In this case, the transport provider has knowledge of the common properties of the signals and performs the appropriate signal transposition in its domain. Of course, the available transposition need not be static and could be a part of the particular service being provided; for example, the transport provider could establish the transposition when the particular service is established. In the following example, illustrated in Figure 6.15, three usage applications that are not collocated need to cooperate. However, since they are not collocated, they require transport facilities in order to cooperate. From a service perspective, such transport facilities are provided by the transport domain, which itself can provide the service by cooperating with other transport domains. This type of peer cooperation is commonly called *federation.*

Each usage domain may communicate with a transport domain having the same characteristic information; in that case, the trails are terminated in the usage domains, which provide their adaptation. The trail termination functions, according to the previous extension, can be compared to an abstraction of several recursive layers. In addition to transport service facilities, the transport service provider may adapt the information delivered by and to each usage domain.

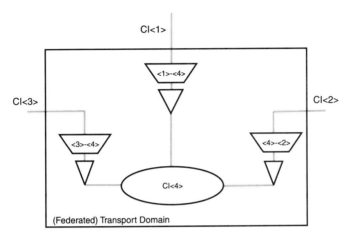

CI - Characteristic Information

Figure 6.15 Complex transport domain.

6.7 General Framework for the Evolution of Switched Networks

In our discussion of shared knowledge in Section 6.3.2, we saw how accidental coupling can be created by sharing more knowledge than is necessary and sufficient and how this can constrain design. Later we saw how a careful choice of shared knowledge of resources across more than one interface can lead to more flexibility. In this section, we will explore how this insight relates to the traditional PSTN as well as to the more advanced intelligent networks. A key point to note is that traditional PSTNs that provide voice services have been strongly influenced by the fact that connection requests operated directly on the network connection performer. This operation was initially very direct, as the dial pulses operated electromechanical switches without any management intervention. While electronic switches later replaced these switches, the architectural principle remained unchanged. Specifically, the service order (dialing) operated directly on the connection performer of the underlying network, and services could only be offered based on the resources of the network (Figure 6.16).

The vision of the intelligent network (IN) was enabled by the concept that service control could be logically separated from connection control. This meant that services could be offered, unconstrained by the shared knowledge of the resources of the underlying network, which allowed the introduction of new added-value services in addition to the basic PSTN. Thus, a service request, such as call transfer, could still be made from a phone terminal, but

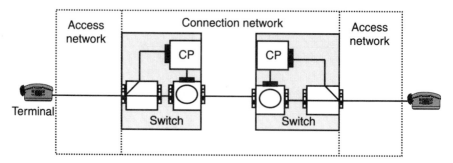

CP : Connection Performer

Figure 6.16 Call control network integration with the connection network.

the request would be delivered to a service manager component rather than being directly delivered to the connection performer. The service manager then becomes a client of the original connection service and can then employ that connection service as needed to carry out the service request. In fact, the service interface can now contain resources that are completely unknown to the original connection service, enabling a wide range of new services to be created without affecting the underlying transport network. We can also consider call control itself to be a service that can also be logically separated from connection control, allowing call control to be processed separately from connection control in the underlying transport network (Figure 6.17).

Note that the notion of considering calls and call control to be a kind of service is only a reflection of the idea that the ability to set up connections in the network may be offered to others. It is not meant to imply that the "call" in ISO usage is no longer a kind of communications session.

In a network that is controlled via management interfaces, a connection set-up request would be processed as shown in Figure 6.18, which shows a model of a network partitioned into two subnetworks. Comparing Figure 6.18 with Figure 6.17, we see that the only difference is that the terminal issuing the service request is no longer the same as the terminal using the service.

This observation leads us to expect that a generic connection set-up service exists that can be implemented with either management or signaling protocols. In fact, the underlying protocols could even be developed outside of the telecommunications environment, since all that is needed is verification that it supports established telecommunications needs (e.g., real time, transaction, security, and reliability). The associated activity can be developed with the viewpoint concepts and methodology we described in Chapter 4, with the selection of the best protocols to be made as part of the engineering process via mapping of the expected QoS parameters (e.g., time for set up) to protocol

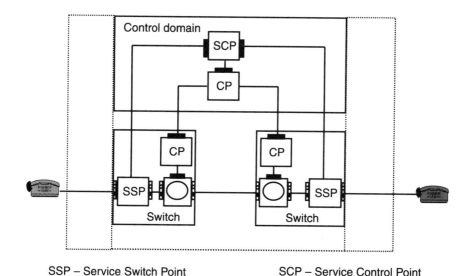

SSP – Service Switch Point   SCP – Service Control Point

Figure 6.17 Separation of service control from connection control—call control in intelligent network architecture.

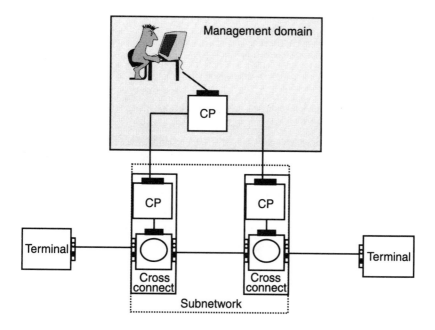

Figure 6.18 Connection control using management interfaces.

characteristics. We can now see how a unified approach can be taken to relate management control of transmission networks and signaling control of switched networks. Specifically, by clarifying the model of how equipment is controlled, we enable specification reuse by separating the computational specification from the engineering specification. Thus, while we may establish different engineering specifications for a service based upon such factors as desired QoS, there is no need for the enterprise or computational specification to change. Further, with some imagination, we can see how we can transform internal management services into commercial usage services by implementing the caller role within a client terminal.

The decomposition of a service activity into interacting service components, each one having its own management interfaces, allows us to define a more general service management approach. Usage of the previously described common generic framework applied to service and service management makes life easier by defining a generic service management for any type of service (such as transport service, switching services, access services, content services, and management services).

6.8 General Framework for the Evolution of Service Access: The TINA-C Network

As seen in the preceding section, the intelligent network was enabled by the separation of service management from connection management. This separation allowed the network operator to provide a new range of services that were no longer constrained by the assumption about the engineering of internal resources of the operator's network. It was recognized that the network would be further enriched if it could be used to deliver third party service, that is, services delivered on behalf of other service providers. To accomplish this, a more general model for service, coupled with a complete separation of service access from service usage, is needed. The establishment of this more general approach to service access has been one of the main results of the Telecommunication Information Networking Architecture Consortium (TINA-C) [14], an organization whose objective has been to drive toward a vision of intelligent networks based upon a homogeneous generalized access platform.

Up to this point, we introduced fundamental and generic concepts. In Chapter 4, we described ODP viewpoint-based specification principles and provided a specific example in Chapter 5. In this chapter, we provided both generic and explicit examples of concepts that increase network agility via progression toward a unified modeling framework. However, general principles and specific applications do not guide the rapid creation of a model for a new service. A framework gives guidance by focusing attention on a reusable set of entities. Just as the functional model of Chapter 3 defines which functions are useful to model transport, a business model provides guidance about which roles are reusable, while a service model provides a pattern of objects and interconnections that can be used over and over for many different service offerings. Such a reuse framework can be applied to realize quick rollout of (almost) any desired service. TINA-C has provided this framework via its business model and service architecture. All the concepts introduced in this chapter, in addition to further clarifying elements, will be used in the following brief exploration of a network embodying the TINA-C architectural principles.

TINA-C defines a general business model [15] that introduces a set of general enterprise viewpoint roles and relations. This model can be seen as a template for an enterprise model at the highest level, whose roles can be specialized and organized into business administrative domains according to specific policies for the business objective of any particular business. Administrative domains can be developed for many purposes (e.g., network monitoring system for management, residential services for business) and can be composed or decomposed to meet domain objectives. A system is then modeled as a set of administrative domains, further elaborated in Chapter 8, and related to each other by business contracts (Figure 6.19).

The business roles are inherently intuitive; consumers order and consume services, retailers sell service on behalf of others, while brokers put consumers, retailers, and providers in touch with each other. In our earlier examples of

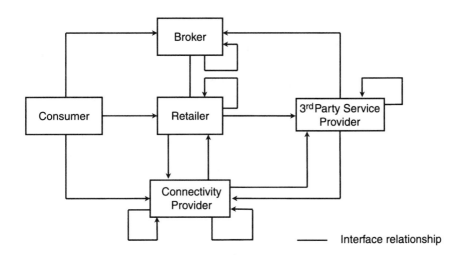

Figure 6.19 TINA-C business model.

the telephone network and the intelligent network, we can easily distinguish the consumer, retailer, and connectivity provider roles. TINA-C adds the notion of brokers and third-party suppliers.

The TINA-C service architecture [16] is based on the separation between service access and service usage. This architecture can be seen as a set of enterprise, information, and computational models that relate service access and service usage to transport network resources. The access part, or session, captures all the interactions between a potential service user and the service provider prior to the service usage. The service usage session allows for the provider to perform service on behalf of the user. Due to the distribution of domains in a TINA-C system, both access and usage sessions need communication facilities that are provided by the connectivity provider role of the business model.

The access session establishes a binding between a service user role and a service provider role (Figure 6.19) and is responsible for establishing a usage session, which will enable fulfillment of the service contract. The usage session will be established with the assistance offered by the connectivity provider role. Of course, the usage session for this service may be an access session for another service; thus, we can construct the recursion in service management that is so essential for general service management. In particular, it is this recursion that allows a retailer to subcontract a part of, or an entire, service to another business role.

If we take the business and service model concepts and apply them to the transport network of Section 6.7, we would have a network that could support third-party business in addition to services provided by the network provider. However, the fixed signal representation of the physical transport network would force consumers and retailers to agree upon the signal representation outside of the network. To be most useful, that network should convey the semantics of any service the consumer and retailer agree to, so the network should make no assumptions about unique signal representation. Figure 6.20 depicts an example network, composed of multiple technologies, transporting a service session between two usage domains. The TINA-C network makes use of the ideas expressed in Section 6.6 to provide for signal representation transposition within the network. This transposition is managed by the adaptation management function alluded to in Chapter 5.

The business model provides a framework for the enterprise model, while the service architecture provides the framework for organizing the communications necessary to access and deliver any service. We have now arrived at a model flexible enough to describe information transfer without limitation to one particular representation as well as a management model whose potential service possibilities are limited only by the imagination of third-party service

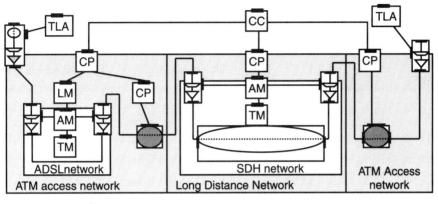

TLA Terminal Layer Adaptation   LM Link Management

CC Connection Coordinator   TM Trail Management

CP Connection Performer   AM Adaptation Management

Figure 6.20 TINA-C network.

providers. In particular, we have a general framework for describing any service together with its means of access and delivery.

6.9 Summary

In Chapter 1, we described the importance of modeling to telecommunications and how it permeates all aspects of design, ranging from service design through equipment and technology design. We also emphasized the need to ensure consistency among these models to fulfill the requirements that govern their development. In particular, we cited the need for a unified approach that would encompass all aspects of networks from planning through operations. We also stressed that a unified approach does not imply a unique model, but several models that can be tailored to meet their specified objectives and always be related each other. This relationship may be achieved via decomposition, specialization, continuity, references, mapping, and/or interactions.

One key objective was to establish a relationship between the transport and management domains, which have generally been treated in a disjoint manner. To facilitate the description of relationships among network functions, we drew upon the generic body of work for describing system functionality. Specifically, in Chapter 3 we presented the functional modeling approach for improving the precision of specification of the transport domain, and in Chapter 4 we introduced a viewpoint specification method for increasing the flexibility and rigor of management domain specifications. In this chapter, we showed how the transport and management domains can be unified by modeling the transport functions introduced in Chapter 3 using the computational objects introduced in Chapter 4. This approach enabled us to fold specification of transport behavior into the same framework as specification of management behavior, enabling a holistic view of the overall network behavior. The rather general treatment of those chapters is then rolled into a more generally useful framework for services in the TINA-C business model and service architecture.

In exploring what knowledge is shared over an interface, we showed how services affect and are affected by equipment and how separation of service access from service usage can increase our ability to offer services that are no longer constrained by assumptions regarding the engineering of the physical network. The examples that we used throughout this chapter do not depend on the selection of a particular distribution of functionality in the network; neither do they depend on the use of a particular protocol to implement the chosen distribution. Thus, the control domain has functionally similar behaviors to the network management domain, though implemented in different ways (e.g., control via SS7 network signaling versus management communications over a DCN). We see that the selection of the component distribution, computing infrastructure, and any protocols is part of an engineering specification process rather than an enterprise specification process. We believe that the concepts described in this book are equally applicable to capturing the relationships with the control model and have shown why we expect this to be valid via our exploration of network evolution.

Thus, we never describe the unified model developed by this approach in its entirety but rather demonstrate how it can be documented via related views, which are connected by the knowledge shared between them. Closing with the story of the blind men and the elephant, each man might come up with a different description of the elephant based upon their individual perspectives, but it is the same elephant!

References

[1]   ITU-T Rec. G.805 "Generic Functional Architecture of Transport Networks," 1995.

[2]   ETS 300 417-1-1 "Generic Requirements of Transport Functionality of Equipment, Generic Processes and Performance," edition 1, Jan. 1996, draft revision, Jan. 1997.

[3]   ETS 300 417-2-1 "Generic Requirements of Transport Functionality of Equipment, SDH and PDH Physical Section Layer Functions," edition 1, Apr. 1997.

[4]   ETS 300 417-3-1 "Generic Requirements of Transport Functionality of Equipment, STM-N Regenerator and Multiplex Section Layer Functions," edition 1, June 1997.

[5] ETS 300 417-4-1 "Generic Requirements of Transport Functionality of Equipment, SDH Path Layer Functions," edition 1, June 1997.

[6] ETS 300 417-5-1 "Generic Requirements of Transport Functionality of Equipment, PDH Path Layer Functions," draft for voting, Dec. 1997.

[7] ETS 300 417-6-1 "Generic Requirements of Transport Functionality of Equipment, Synchronization Distribution Layer Functions," final draft after Public Enquiry, Jan. 1998.

[8] ETS 300 417-7-1 "Generic Requirements of Transport Functionality of Equipment, Equipment Management and Auxiliary Layer Functions," draft version 0.5, 10 July 1998.

[9] ISO/IEC 10746-1 / ITU-T Rec. X.901 "Information Technology, Open Distributed Processing, Reference Model: Overview," 1997.

[10] ITU-T Rec. G.851-01 "Management of the Transport Network, Application of the RM-ODP Framework," 1996.

[11] ITU-T Rec. G.852-01 "Management of the Transport Network, Enterprise Viewpoint for Simple Subnetwork Connection Management," 1996.

[12] ITU-T Rec. G.853-01 "Common Elements of the Information Viewpoint for the Management of a Transport Network," 1996.

[13] ITU-T Rec. G.854-01 "Management of the Transport Network, Computational Interfaces for Basic Transport Network Model," 1996.

[14] TINA-C, the "Telecommunications Information Networking Architecture Consortium" at http://www.tinac.com/.

[15] TINA-C *Business Model and Reference Points,* Version 4.0, May 1997, available from http://www.tinac.com/97/bm_rp.pdf.

[16] TINA-C *Service Architecture, Version 5.0,* June 1997, available from http://www.tinac.com/97/sa50-main.pdf.

7

Usage of Unified Modeling Language

7.1 Introduction

By 1993, the two major object-oriented methods, the Rumbaugh [1] and Booch [2] approaches, had converged to the point that they had become more similar than different, with remaining differences related more to terminology and notation than semantics. Further, based upon increasing experience in applying object-oriented methods, it had become possible to distinguish between fundamental concepts of primary importance and those of lesser importance. Thus, at the end of 1994, Jim Rumbaugh and Grady Booch decided to unify and enhance their respective approaches into a single unified method. During the initial development of this method, the emphasis shifted from the method to the creation of a universal language for object-oriented modeling and "The Unified Modeling Language for Object-Oriented Development" was born. UML was developed by Grady Booch, Ivar Jacobson, and Jim Rumbaugh, from Rational Software Corporation, together with many contributions from other methodology and software experts, vendors, and users. Version 1.0 of UML was provided to the Object Management Group (OMG) in January 1997 for standardization, and by August 1997 Version 1.1 of UML was adopted [3–4]. This chapter will describe the basic concepts of UML and, by example, its application to the telecommunications domain. The introduction to UML provided hereafter is limited to the subset of constructs that is relevant to our immediate purpose, that is, ODP-based specification. The UML notation [5] is much more powerful than that which is described here.

7.2 Overview of UML Modeling Concepts and Language

UML can be applied to all phases of the life cycle of a system, both hardware and software, encompassing those of functional requirements capture, analysis and design, and ultimately implementation. The following subsections illustrate how this may be accomplished.

7.2.1 Functional Requirements Capture

In UML, the functional requirements of a system are captured from the point of view of roles that make some usage of the system. These roles and their relationship to types of usage of the system are analyzed one at a time, and each type of usage is called a use case. This view of the system from each potential user is very useful because different users have clearly different and complementary perspectives. For example, in a trading post, there are importer roles (those who buy goods) and exporter roles (those who sell goods), each of which have their own requirements. The complete set of system use cases enables a comprehensive description of its requirements from the standpoint of all of the potential roles, where we note that a role may be played by external systems or human persons. Since a human user or system may have to play several roles, the relationships among the roles, use cases, and graphical representation are necessarily decoupled. Given that a system manager is also a potential user, we represent management policies as use cases as well. The complete set of use cases very closely represents a subset of what we defined in Chapter 4 as an enterprise specification. Networks can also be viewed from the perspectives of different roles; for example, a service invoker role is interested in whether service has been successfully delivered or not, while a notification receiver role is interested in receiving indications of system faults, and a maintenance role is interested in initiating system repair activities. Use case diagrams show the various required uses (or utilization) of the system of interest, describing how the system may interact with its environment and the set of roles that are interacting with the system, which are called actors. Each actor operating on the system of interest represents a particular perspective. The use case diagram, illustrated in Figure 7.1, shows which actors interact with each use case and how use cases may in themselves be composed of other use cases. The actual specification of the use case may rely upon any of the behavior specification techniques to be described later, including plain textual descriptions (as were used for describing enterprise behavior in Chapter 4).

7.2.2 Logical Architecture

The logical architecture of the system of interest is, as expected, defined in terms of objects that interact dynamically by exchanging messages to realize

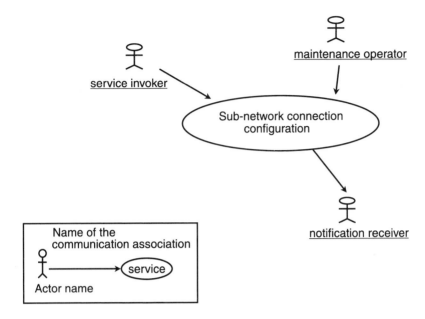

Figure 7.1 Example use case diagram.

the behavior defined by the use cases. Object interactions are depicted in sequence diagrams that may be elaborated by developing associated scenarios, which sequentially step through various instances of a use case. The relationships identified between interacting objects are shown in a collaboration diagram, and class relationships are shown in a class diagram.

Packages[1] are used to group object classes and their defined relationships, as illustrated in Figure 7.2. Their purpose is much like that of the information model fragments described in Chapter 3, which represent managed object classes and relationships that have been grouped according to their pertinence to some gross aspect of the system (i.e., equipment fragment and performance fragment). Additionally, packages may be contained in other packages, which gives us the ability to completely and clearly organize the system specification and to reuse fragments of specifications.

7.2.2.1 Object Classes

Class diagrams depict the object classes that model the entities of the system and their associated interobject relationships. Interface definitions, as well as code and/or data base schema, may be generated from object class specifications.

---

1. A UML package is not the same as a GDMO package, which is a collection of commonly occurring groups of attributes, actions, notifications, and behaviors that can then be used in object definitions.

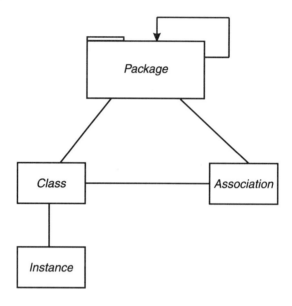

Figure 7.2 E-R representation of logical architecture modeling elements.

From a diagrammatic perspective, UML represents object classes in terms of a named rectangle providing the class name, and when necessary, the class attributes and operations (Figure 7.3).

7.2.2.2 Relationships

Relationships between object classes are called *associations* in UML, as illustrated in Figure 7.4. The cardinality of the association can be attached to roles of an

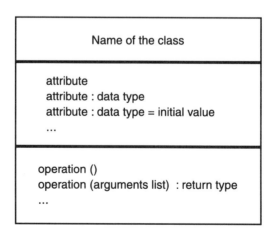

Figure 7.3 Object class representation.

association and is called a *multiplicity*. For those accustomed to other notation (e.g., entity-relationship diagrams), it can be confusing to see the multiplicity placed close to the target class; for example, in Figure 7.4, Class 1 objects are related to zero-to-many objects of Class 2, and Class 2 objects are related to at most one object of Class 1. In UML, we define an association, which can be named or unnamed, and where the involved roles can be played by different classes. An *aggregation* represents a special kind of association in which one of the object classes assumes the role of a *container* and the other one represents a *contained element*.

As illustrated in Figure 7.5, associations can be declared as nonnavigable, navigable in one direction, or navigable in both directions (which is the default value in UML). Specifying the navigability of an association in the data model provides the potential for navigating between objects of the classes related by that association. Also, during the code generation phase, while nonnavigable specifications generate smaller size code, bidirectional association results in larger size code resulting from the two cross-referencing pointer attributes (one in each object class).

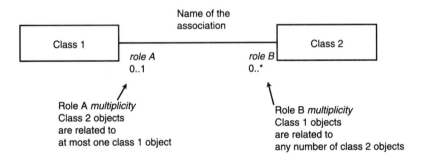

Figure 7.4 Representation of associations.

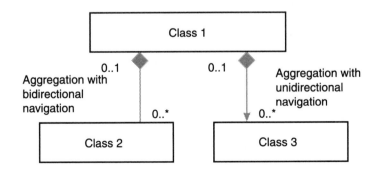

Figure 7.5 Navigability in associations.

When the associations between object classes themselves have attribute(s) and/or operation(s), UML enables the capture of these characteristics as an association class. We can explain this concept using the example of a school, where we find forms and classrooms. A "classroom" object class is characterized by its location and number of seats, whereas a "form" object class is defined by the year and number of students. An association defined between these two classes could have the name "occupancy" and have two attributes: the start and end times of occupancy of the classroom according to form. Clearly, neither of these two attributes could be attached to the form nor classroom object classes. In UML, we define an association class to represent the occupancy, which can be named or unnamed and where the involved roles can be played by different classes (Figure 7.6). The associations between object classes define the potential for interactions between objects of these classes. Generally, groups of objects cooperate with each other to provide a more complex service than the services provided by each object individually. The knowledge of which objects can interact with a given object is defined by the associations between them.

As mentioned in Chapter 2, *generalization* (the ability to organize our classes into a hierarchy) is a major principle of the object-oriented paradigm. This is called *inheritance* in languages like C++ and represents the relationship between object classes. In UML, and as depicted in Figure 7.7, generalization is represented by an arrow from the subtype to its super-type.

7.2.2.3 States and Behaviors

UML state diagrams describe possible states of objects, possible transitions between these states, events that trigger the transitions, and consequent actions that may result. As illustrated in Figure 7.8, UML represents state in terms of a rounded-corner rectangle containing the name of the state, entry action,

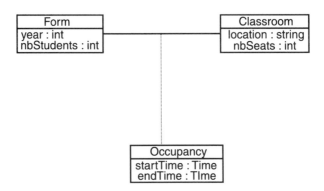

Figure 7.6 Capturing characteristics of an association.

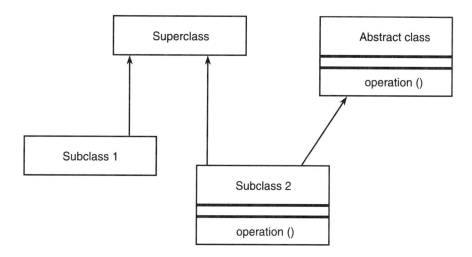

Figure 7.7 Representation of generalization.

Figure 7.8 Representation of state.

associated activity, and consequent actions resulting from the occurrence of particular events. An activity is a graph of actions, whereas actions are atomic "functions" or "procedures."

Composite states of a system may also be graphically depicted, as shown in Figure 7.9. Just as functions help us structure software, a composite state is useful in helping structure the overall state diagram. This can facilitate the reuse of state machine specifications and assist in avoiding "clutter" at the current level of abstraction.

As mentioned earlier, object interactions define the behavior of the system. In UML, these interactions are captured in terms of *sequence diagrams* and *collaboration diagrams.* The sequence diagram, illustrated in Figure 7.10, focuses upon showing the time order of messages and the object life times involved in invoking operations between objects.

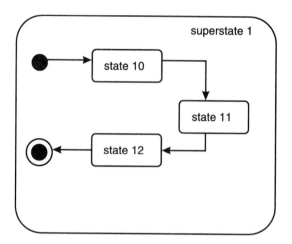

Figure 7.9 Representation of composite states.

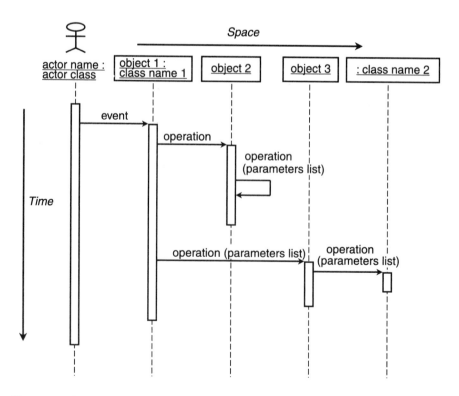

Figure 7.10 Sequence diagram capture of object interactions.

Collaboration diagrams provide another way to represent object interactions, focusing upon object associations rather than the time order of messages (Figure 7.11). These diagrams cannot depict the behavior of a group of objects in its entirety, as this would be too difficult to represent graphically. However, they provide a convenient means for representing a limited number of scenarios of cooperation inside the system or between the system and its environment. When describing such scenarios, it is not always necessary to identify the object instance when the necessary information is contained in the class of the object (e.g., "postal clerk"). In some cases it is not even necessary to identify the class when the object is the single instance of the class and is well known in the whole system (e.g., "president").

This leads us to the observation that collaboration and sequence diagrams do not necessarily rely upon class information and so can be used to depict interactions before classes have been defined. For example, they can be used to depict high-level interactions in an organization, or between different organizations, when the messages between the objects represent data flows.

In UML, three levels of behavioral specification are possible.

1. Specification of the behavior of a group of objects (i.e., a system or subsystem): sequence and collaboration diagrams are most suitable for this purpose.

2. Specification of the behavior of an individual object: state chart diagrams provide the best means for defining object states and transitions.

3. Specification of the behavior internal to an operation (i.e., what happens when a message is received by an object that triggers the performing of the corresponding operation): this behavior is best described by activity diagrams that show which elementary actions are performed, how they are organized, and the final state of the object at the end

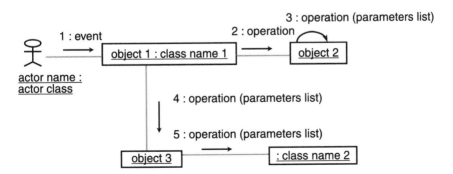

Figure 7.11 Collaboration diagram capture of object interactions.

of the activity. Since activity diagrams are the most complete in terms of representation power (e.g., they provide the means to represent various kinds of synchronization), they are often used to model the behavior of complete systems (i.e., group of objects) or in business process modeling.

7.2.3 Extensibility Mechanisms

In UML, three extensibility mechanisms are available to the end-user: *stereotypes, tagged values,* and *constraints.* Stereotypes are the most important of these extension mechanisms and provide a means for defining new "virtual" constructs in the UML language (e.g., classes, relationships). Therefore, when writing a UML model, the designer can specify extensions to the UML language itself, which may encompass new values, additional constraints, or new graphical representations. By allowing stereotypes to have associated graphical representations, users can introduce new ways of graphically distinguishing model elements classified by a particular stereotype. As an illustration, using the stereotype extensibility mechanism, the designer can define the *InformationObjectType* class as a kind of UML class. The designer is then able to define the information viewpoint object class *SubnetworkConnection* by instantiating the *InformationObjectType,* capturing the appropriate semantics as explained in Chapter 4. We note that while both the *SubnetworkConnection* and *InformationObjectType* definitions are captured in the same UML specification they do not address the same application (i.e., the first being an object class representing a transport network entity, while the second captures the semantics of an ODP concept). As a result, the *SubnetworkConnection* object class would be an instance of the *InformationObjectType* class, which itself is defined as a stereotype of the UML class concept. As another example, it is valuable to define *EnterpriseSpecification* as a stereotype of the UML model class. Of particular interest for future work is the ability to attach new graphical representations to a stereotype. This opens the path toward using UML to express the functional model constructs of Chapter 3 and their unification with specific computational objects, as discussed in Chapter 6.

Tagged values enable us to add properties to UML constructs where it is not possible to add attributes. A tagged value is characterized by its name and its value (as it has no type). The value, represented as a character string, is not part of the UML language and cannot be checked by the UML tool set. The value is, however, available to other tools such as code generators or more domain-specific semantics checkers. We will not pursue the discussion of extensibility mechanisms further in this chapter.

7.2.4 Object Constraint Language

The object constraint language (OCL) is a small language (in the sense that the number of constructs of the language is small) that complements UML. OCL provides a means to specify functional integrity constraints in a simple way and can be used at any layer of the model architecture. Thus, it makes it possible to express such integrity constraints that cannot be expressed using only UML, such as "a tandem connection is a sequence of link connections and subnetwork connections, where it is not possible to have two contiguous link connections nor two contiguous subnetwork connections." OCL is often considered as a simplified Z notation[2] [6]. In spite of OCL being simpler, in most cases a plain text explanation of the OCL representation is also attached to the object.

7.3 Relating UML and Network Management Domain Specifications

As mentioned in Chapter 5, UML concepts and notation can be used to capture RM-ODP-based network management domain specifications (i.e., the G.851-01 meta-model[3] [7]). Currently, standard object models for the management of transport networks primarily employ UML use case diagrams and class diagrams, though state diagrams could be provided as needed. Table 7.1 provides a list of the UML concepts that are relevant to network management domain viewpoint specifications.

We first illustrate how UML can be used for modeling the G.851-01 approach itself (this is captured in what we call the G.851-01 meta-model), essentially using class diagrams. Then, starting from the concepts provided in Table 7.1, we show which UML constructs are relevant to modeling network management services (such as the topology management service).

7.3.1 The G.851-01 Meta-Model Expressed in UML

This section illustrates how UML class diagrams can be used to represent a viewpoint-based architecture, interviewpoint relationships, and the modeling concepts that are relevant to each viewpoint. As a reminder:

---

2. The Z notation [4] allows for a formal description of an information system by using mathematics to fully describe the system. In particular, Z uses the mathematical data type to model the data in the system together with predicate logic to describe the effect of each operation. These concepts are not oriented toward any particular computer representation but both concepts have a rich set of mathematical laws that make it possible to reason about the system. Perhaps not intuitive is the fact that a mathematical function is a data type and can be used as such in specifications. .

3. A meta-model is defined as a model that defines the language for expressing a model.

Table 7.1
UML Concepts Relevant to Network Management Domain Viewpoint Specifications

Viewpoint	UML Diagram	Relevant UML Concept
Enterprise	Use case diagram	Actor, use case
Information	Class diagram	Class, association, attribute, role,
	State chart diagram	multiplicity, state, transition
Computational	Class diagram	Operation, interface, parameter, exception
Engineering	Class diagram	Class, attribute, interface, operation, parameter, exception

- Enterprise viewpoint modeling concepts include community, roles, actions, and policies.
- Information viewpoint concepts include information object classes, attributes, interclass relationships, permitted states of the system and/or any subsystem, and permitted transitions.
- Computational viewpoint concepts include computational object types, interfaces, operations, parameters, and exceptions.
- Engineering viewpoint concepts include engineering object types, interfaces, operations and actions, parameters, and exceptions.

In order to avoid confusion, the reader should keep in mind that this meta-model could have been expressed in a traditional entity-relationship notation.

An RM-ODP-based management specification (e.g., G.851-01 compliant as discussed in Chapter 4) can be modeled as a UML package. This package is composed of four other packages corresponding to the enterprise, information, computational, and engineering viewpoint specifications, respectively. In turn, each of these packages is composed of a number of modeling constructs. Figure 7.12 shows the overall architecture in terms of viewpoint specifications and how each viewpoint specification is associated with a UML package.

7.3.1.1 Enterprise Specification

The various entities of an enterprise specification are expressed in UML by object classes, and the relationships between them are represented by UML associations. As depicted in Figure 7.13, an enterprise specification is made up of enterprise roles, enterprise policies, and enterprise actions, for example. An enterprise policy is attached to either the whole community (see, e.g., the association between the class EnterprisePolicy and the class EnterpriseSpecifica-

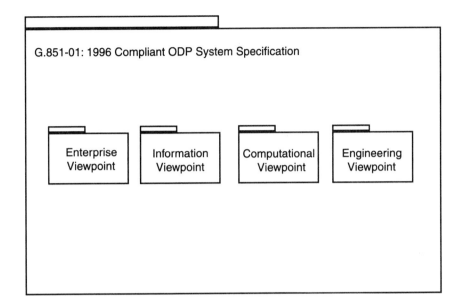

Figure 7.12 Association of ODP system specification with viewpoint packages.

tion) or to a particular action (association between the class EnterprisePolicy and the class EnterpriseAction). Multiplicities (the UML name for cardinalities) make it clear that an enterprise specification (class EnterpriseSpecification) may contain many enterprise roles (class EnterpriseRole), while each enterprise role is related to at most one single enterprise specification; the cardinality of the other components is also indicated. This description of the G.851-01 meta-model expressed in UML is much more rigorous than the existing one contained in ITU-T Recommendation G.851-01.

For most of these G.851-01 meta-model constructs, we create stereotypes that enable the establishment of precise specifications; indeed, they are more precise than if we used standard UML meta-model constructs. As an illustration, we create the stereotype <<EnterpriseRole>> from the UML actor meta-class and the stereotype <<EnterpriseAction>> from the UML use case meta-class. Enterprise roles and actions are then naturally mapped onto these stereotypes, which then serve as a basis for specifying the topology management service, the result of which is shown in Figure 7.14. This will be detailed in Section 7.3.2.1.

7.3.1.2 Information Specification

The G.851-01 meta-model constructs used to express an information viewpoint specification are illustrated in Figure 7.15. This figure specifies the elements needed to provide a well-formed G.851-01 information specification. In the

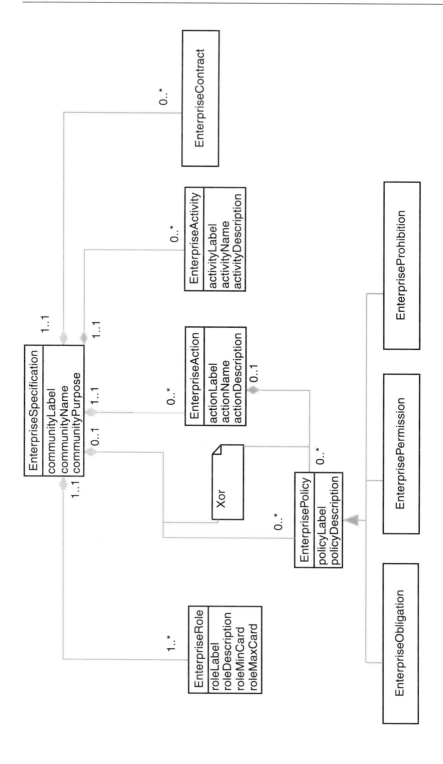

Figure 7.13 G.851-01 enterprise viewpoint meta-model.

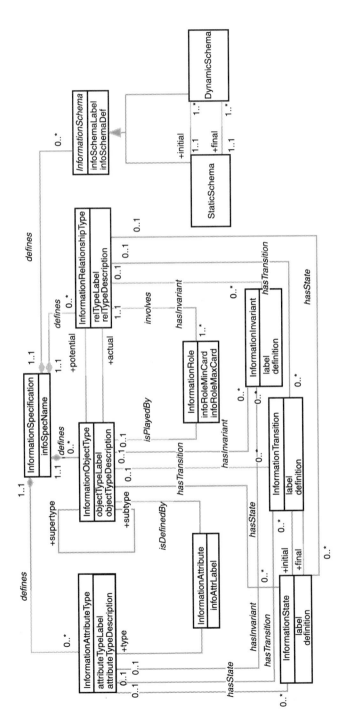

Figure 7.14 UML topology management enterprise specification.

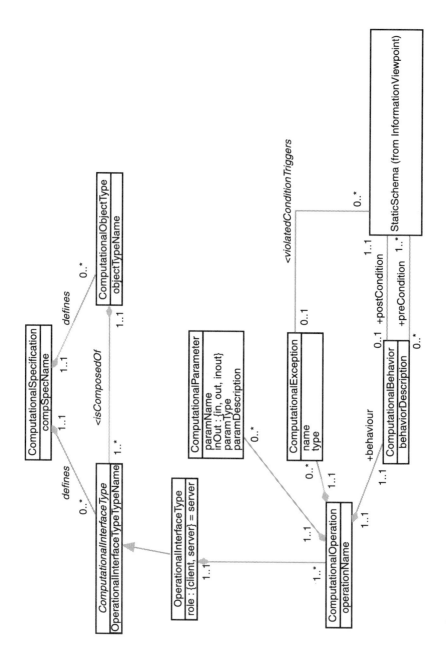

Figure 7.15 G.851-01 information viewpoint meta-model.

information viewpoint, a key modeling concept is the information object type, which is important enough for us to create a stereotype called <<Information ObjectType>>. We also create a stereotype <<InformationRelationship Type>>. An actual information specification makes use of these stereotypes, as illustrated in Figures 7.16 to 7.18, for the topology management service information viewpoint specification in UML. These figures are explained in Section 7.3.2.2.

7.3.1.3 Computational Specification

Similarly, the G.851-01 constructs used to express a computational viewpoint specification are illustrated in Figure 7.19. A well-formed computational specification (class ComputationalSpecification) is composed of computational object type (class ComputationalObjectType) and computational interface type (class ComputationalInterfaceType) definitions. A computational object type is a collection of computational interface types and an operational interface type (class OperationalInterfaceType) is a kind of computational interface type. Just as for the other viewpoints, we create stereotypes to support the most important

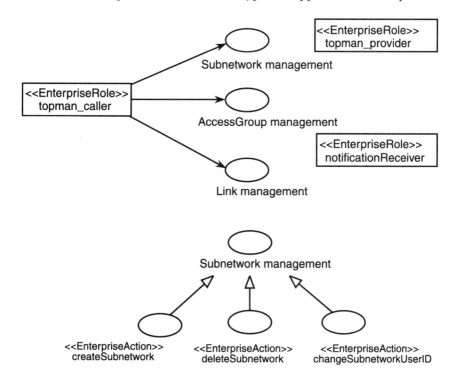

Figure 7.16 UML topology management information specification—generalization (inheritance).

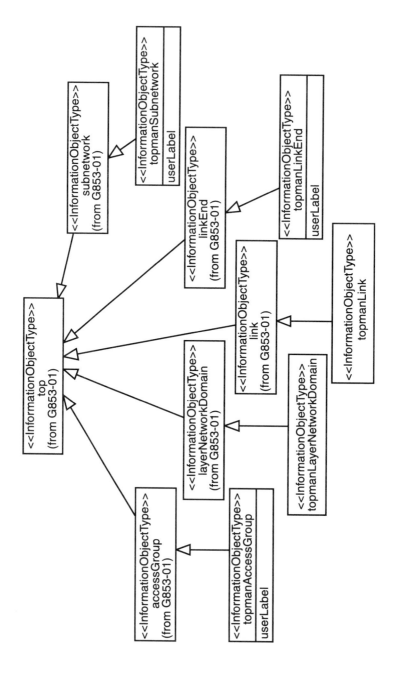

Figure 7.17 UML topology management information specification—a sample association.

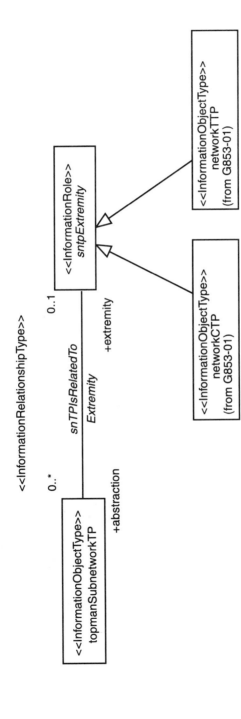

Figure 7.18 UML topology management information specification—example association.

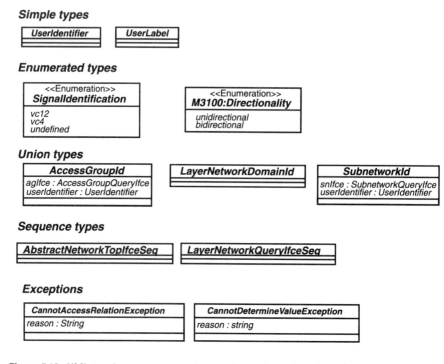

Figure 7.19 UML topology management computational specification—base parameter types.

concepts. Since the G.85x series does not standardize computational objects, we have no need to define a stereotype for this modeling concept.

G.851-01 computational viewpoint interfaces are mapped directly onto the <<OperationalInterfaceType>>, which is a stereotype of the UML interface type. G.851-01 computational operations (class ComputationalOperation) map directly on to UML operations; parameters (class ComputationalParameter) are mapped onto UML types.

Figures 7.20 and 7.21 will illustrate the use of these stereotypes in the case of the topology management service, to be described in Section 7.3.2.3.

7.3.2   Topology Management Service Modeled in UML

We will now detail how to apply the meta-model of G.851-01 in UML (which serves as a generic template) to the specific example of the topology management service, which was provided in Chapter 5.

7.3.2.1   Enterprise Viewpoint Specification—Topology Management Service

The enterprise viewpoint uses the UseCase diagram to describe the different actions defined in the service (Figure 7.14). The topman_caller,

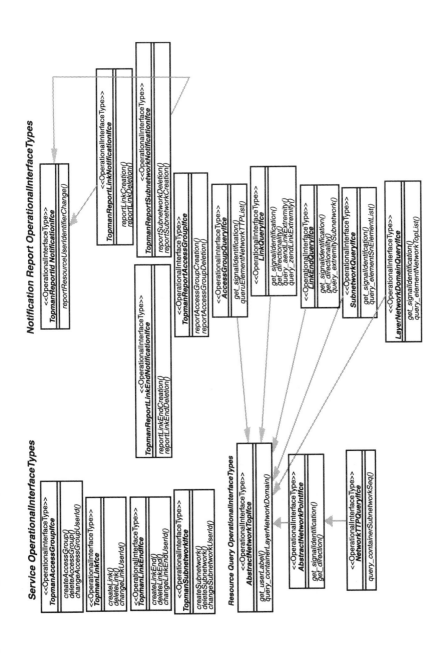

Figure 7.20 UML topology management computational specification—example interface types.

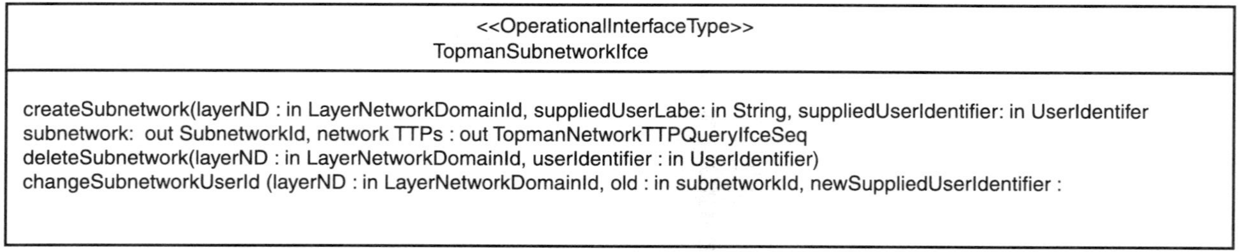

<<OperationalInterfaceType>>
TopmanSubnetworkIfce

createSubnetwork(layerND : in LayerNetworkDomainId, suppliedUserLabe: in String, suppliedUserIdentifier: in UserIdentifer
subnetwork: out SubnetworkId, network TTPs : out TopmanNetworkTTPQueryIfceSeq
deleteSubnetwork(layerND : in LayerNetworkDomainId, userIdentifier : in UserIdentifier)
changeSubnetworkUserId (layerND : in LayerNetworkDomainId, old : in subnetworkId, newSuppliedUserIdentifier :

Figure 7.21 UML topology management computational specification—TopmanSubnetworkIfce operations and signatures.

topman_provider, and the notificationReceiver are defined not simply as UML actors but more precisely as <<EnterpriseRole>>, a stereotype of actor, as prescribed in Section 7.3.1.1. Similarly, the createSubnetwork, deleteSubnetwork, and changeSubnetworkUserId actions are defined not simply as use cases but as <<EnterpriseAction>>, a stereotype of use case. This is a very good illustration of how the UML stereotype extension mechanism can help a designer to develop more precise specifications. Once the extensions have been defined, the writer of a G.851-01 compliant specification can continue design of the specification without the need for knowledge of standard UML notation.

7.3.2.2 Information Viewpoint Specification—Topology Management Service

We now apply the stereotypes identified in Section 7.3.1.2 to the topology management service example. Figure 7.16 illustrates the overall generalization hierarchy where each information object class is defined as having the stereotype <<InformationObjectType>>. Again, the specifier does not need to manipulate pure UML classes but rather the information object types that are of more relevance to a Rec. G.851-01 specialist. Figure 7.17 shows an example of inheritance between information object types defined in different packages: the topmanLinkEnd, defined in the topman service package, inherits from linkEnd defined in the Rec. G.853-01 package.

Similarly, Figure 7.18 shows an example of an association defined in the topman service information specification. The association is designated <<InformationRelationshipType>>, and interrelated object classes have the stereotype <<InformationObjectType>>. One of the two roles, namely sntpExtremity, of that association has the stereotype <<InformationRole>>. The two information object types that inherit from sntpExtremity are the networkCTP and networkTTP types. This provides a convenient means for modeling the fact that a given association role may be played by a number of different object classes, which cannot be modeled by the standard UML notation.

7.3.2.3 Computational Viewpoint Specification—Topology Management Service

We now apply the modeling guidelines identified in Section 7.3.1.3 to the computational viewpoint of the topology management service example. In this case, standard UML types provide a good fit for representing parameter types in the operation signatures (Figures 7.19 and 7.20).

We can distinguish service, notification (report), and query interfaces (viewing) computational interfaces as follows:

- Service interface types gather operations that provide the actual service.

- Query interface types are made up of operations that enable retrieval of data accessible through these interfaces.

- Notification interface types contain the types of event notifications that can be issued by the objects supporting these interfaces.

Figure 7.20 presents an overview of the different interface types defined for topology management, which all have the stereotype <<OperationalInterfaceType>>. For the sake of clarity of the diagram, we only present the list of operations that comprise part of each interface type (the signature definition is not shown). We note that inheritance between interface types has the same meaning as between object classes. Specifically, the total set properties available in a subclass is composed of the set of properties defined locally (i.e., in the subclass) in addition to those inherited from the super-class(es). For example, as shown in Figure 7.20, operational interface types may exist between AccessGroupQueryIfce and AbstractNetworkTopIfce. The associated semantics are that the total set of operations accessible through the computational interface type AccessGroupQueryIfce encompasses:

- The computational operations defined locally (i.e., get_signalIdentification() and queryElementNetworkTTPList());

- Those inherited (i.e., get_userLabel() and query_containerLayerNet workDomain()).

In Figure 7.21, we zoom in on the TopmanSubnetworkIfce and show each operation separately with its signature. This is the collection of three computational operations: createSubnetwork, deleteSubnetwork, and change-SubnetworkUserId. For each of these operations, the parameters are defined by a name (e.g., layerND), a passing mode (e.g., in[4]), and a type (e.g., LayerNetworkDomainId).

---

4. In the meta-model, ParameterDirectionKind defines an enumeration whose values are *in, inout, out,* and *return*; these values connote if a parameter is used for supplying an argument and/or for returning a value. Specifically:
   - *In:* parameter has to be given a value by the client upon invoking the operation;
   - *Out:* parameter is given a value as the result of the provider performing the operation and passing the value back to the client;
   - *Inout:* parameter is given a value by the client; this value can be changed by the provider and passed back to the client.

7.4 Summary

In this chapter, we provided an overview of the most important UML constructs that may be used to express the RM-ODP-based management specifications described in Chapter 4. We note that UML has the means to reinforce viewpoint linkages, providing a tight binding between the information and computational viewpoints, in particular. In the remainder of the book, we will use UML for expressing RM-ODP viewpoint specifications.

References

[1] Rumbaugh, J., et al., "Object-Oriented Modeling and Design," Englewood Cliffs, NJ: Prentice-Hall, 1991.

[2] Booch, G., "Object-Oriented Analysis and Design with Applications," 1st ed., Redwood City, CA: Benjamin/Cummings, 1991.

[3] UML 1.1: Semantics, Object Management Group, Framingham, MA, 1998.

[4] UML 1.1: Semantics, Object Management Group, Framingham, MA, 1998.

[5] Rumbaugh, J., I. Jacobson, and G. Booch, "Unified Modeling Language Reference Manual," Reading, MA: Addison-Wesley, 1998.

[6] Spivey, J. M., *The Z Notation, A Reference Manual,* 2d ed., Prentice-Hall International Series in Computer Science.

[7] ITU-T Rec. G.851-01, "Management of the Transport Network, Application of the RM-ODP Framework," 1996.

8

Interdomain Management

8.1 Introduction

Requirements and constraints on today's large communications networks are changing rapidly. The introduction of new services—such as universal personal communications, multimedia, and broadband service—requires sophisticated networking and management features and a large degree of flexibility from the networks that support them. Accommodating requirements from emergent services, in the context of the existing networking infrastructure and network and services management base, is a major challenge.

In very large networks, there may be dozens of applications and services with between thousands and millions of associated objects representing resources that have to be administered and controlled. This diversity of services and networks introduces a range of challenges and requirements—with respect to cost, complexity, performance, and scalability—for both network and service management solutions. Deploying a single management approach to address this huge array of complex management tasks, with their associated diverse requirements and constraints, is not feasible. A better solution is to structure and partition management responsibilities according to a set of management domains distinguished by specific characteristics, and apply management policies and technologies tailored to these characteristics on a per domain basis. Such a partitioning philosophy also lends itself to following natural divisions that may arise from network operator business model, security, and administrative policies. We note that other motivators for subdivision of overall network infrastructure and network and service management domains can include facilitating decentralized control of databases, network configuration, and resources. Therefore, distinct domains naturally arise as the result of the inherent complexity of management applications and networks. Further, whenever new management applications and protocols are introduced into a large, complex network, it is necessary to assure that they peacefully coexist and seamlessly interwork with existing applications. Thus, management domains may also be created based upon technological (e.g., CMIP, SNMP, CORBA, DCOM, Java) drivers. Such partitioning of management functionality and deployment of management domains drives the need for interdomain management solutions.

In this chapter, we introduce domain concepts and types, identify some key issues associated with interdomain management, and provide a more complete treatment of interdomain management when the domains are characterized by the CMIP, SNMP, or CORBA management technologies. Interdomain management among these three technologies is particularly relevant because CMIP and SNMP are widely deployed in current management systems, and CORBA is emerging as a standard for distributed object computing and systems integration that can also be used for the design and deployment of network and service management systems.

8.2 Domains

The concept of domain is not unique to network and service management. In this section, we will examine the general concept of domain from the perspectives of both systems engineering and computer science to establish a broader understanding of what domains are, how they arise, and how they may be characterized. Domains represent a powerful modeling concept that can simplify the analysis and description of complex systems. There are many types of domains (e.g., management domains, naming domains, language domains, and technology domains).

In mathematics, the theory of functions defines a domain as the set of argument values for which a function is defined. *Domain theory* is a branch of mathematics, introduced by Dana Scott and Christopher Strachey in 1970 [1], as a mathematical theory of programming languages. For nearly a quarter of a century it was developed almost exclusively in connection with denotational semantics in computer science. In this context, a domain is a mathematical structure consisting of a set of values and an ordering relation on those values. Domain theory is the study of such structures.

With respect to the Internet, "domain" is most commonly used to refer to a group of computers whose hostnames share a common suffix, the domain name. Some important domains include .com (commercial), .edu (educational, mostly in the United States), .net (network operations), .gov (U.S. government), and .mil (U.S. military). Most countries also have a domain, for example, .us (United States), .es (Spain), and .fr (France).

In object computing, a domain represents an enclosing scope for a collection of objects (members of the domain) that are associated with some common characteristic. A domain can itself be modeled as an object and may also be a member of other domains [2]. The ODP domain definition contained in Rec. X.902 [3] is: "A set of objects, each of which is related by a characterizing relationship <X> to a controlling object." According to this definition, every domain has a controlling object associated with it, and it is the relationship that each object of the domain maintains with the controller object that defines the domain scope. While domains allow partitioning of systems into collections of objects, we note that it is the object associations, or bindings defined among them, that characterize a domain. The ODP definition of domain also does not exclude the possibility that an object may be a member of several domains (which may be of different kinds), so the sets of members of domains may overlap. Examples of domains, according to the common characteristic that defines the members of the domain, include:

- *Referencing* domain: the scope of an object reference;
- *Representation* domain: the scope of a message transfer syntax and protocol;
- *Security* domain: the extent of a particular security policy;
- *Type* domain: the scope of a particular type identifier;
- *Transaction* domain: the scope of a given transaction service.

In the context of networking, we usually define an administrative domain as a collection of hosts, applications, and interconnecting network(s) that are managed by a single administrative authority. Technology domains are usually identified in terms of common protocols, syntaxes, and similar build-time characteristics. Management domains are generally either administrative or technological in nature. For example, two telecommunications management networks associated with two different public network operators represent two separate administrative domains; on the other hand, two network management systems associated with the same public network operator, where one employs OSI systems management and the other employs CORBA, represent two separate technology domains. These concepts are illustrated in Figure 8.1.

Domains may be related in many ways—for example, *containment,* where a domain is contained within another domain, and *federation,* where two domains are joined in a manner agreed to and set up by their administrators. The federation approach to interdomain management is especially attractive

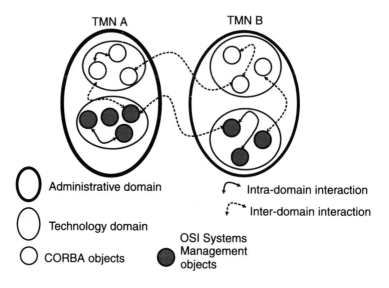

Figure 8.1 Administrative and technology domains.

when we consider the fact that management applications in separate domains are usually designed, developed, and deployed independently, without any previous or future coordination.

8.3 Interdomain Management Issues

As discussed in Chapter 4, the ODP framework of abstractions provides a modeling technique that supports the ability to specify network behaviors and logical object-oriented software architectures independent of their physical distribution. In this section, we discuss the issues that may arise when such behaviors are implemented in real systems.

From a computational viewpoint perspective, an ODP system may be visualized as a set of computational objects that interact with each other, although they may be remote from one another. Thus, computational object interactions are represented independent of their actual implementation in real systems (i.e., they share a uniform view of the underlying heterogeneous distributed system). However, from an engineering viewpoint perspective, an ODP system is not wholly uniform. Not only may objects reside in different physical locations, there may be many other instances of heterogeneity across the distributed system, including those related to processor architecture, networking mechanisms, data representations, and security policies. Thus, distributed computing infrastructures are needed to provide the execution environment and support the realization of distribution transparencies. There are a number of issues relevant to interdomain scenarios that arise due to the transition from the computational to the engineering viewpoint. The design of interdomain solutions must carefully consider heterogeneity in order to achieve the desired end-to-end interactions.

8.3.1 Interoperability and Interworking

A common issue in technological domains is related to the possibility of heterogeneity or homogeneity in the interface models used at an interdomain boundary. For example, one domain can be based on CORBA technology, while the other employs OSI systems management. When comparing CORBA and OSI systems management in the computational viewpoint, there are some similarities and some important differences. An interdomain solution in such a scenario must provide all the mechanisms necessary to ensure that objects in each domain can work together.

We can distinguish two types of interdomain management scenarios, based upon the heterogeneity of their associated computational viewpoint models:

- *Interoperability* involves the use of the same computational model at the interdomain boundary. Here, interdomain solutions are provided by means of bridges.

- *Interworking* involves different computational models at the interdomain boundary. Here, interdomain solutions are provided by means of gateways, which must compensate as much as possible for the differences between the computational models for the management domains being connected.

We note that this categorization simply refines the commonly held understanding of the term *interoperability,* which is used to express the need for entities (objects, components, systems, networks) providing different implementations of some functionality, to work together according to some expected behavior.

The difficulty of designing and deploying interworking solutions can range from relatively easy to extremely difficult. Relatively simple interworking solutions are possible when the constructs in both models—such as calling conventions and data types—are more or less semantically equivalent (the models are almost semantically isomorphic), and a reasonable level of interworking between the two domains can be achieved through straightforward mappings. Difficulties arising from mismatches are relatively easy to resolve when they occur in the engineering viewpoint but are more problematic when the mismatch occurs in the computational viewpoint and still even more difficult if they occur at the information viewpoint level. The latter case usually results in very complex interworking solutions, which may involve conversion and agreement upon common information schemas.

The usage of the terms *bridge* and *gateway* is purely conceptual and refers only to the functionality required to perform the interdomain mapping. The actual implementation and location of such functionality is open to several alternatives.

8.3.2 Interdomain Mapping Functionality

The goal of any particular deployment of interdomain functionality is to support the seamless interoperation of objects running in distinct domains. Interdomain mapping provides the functions required to make interdomain interactions possible by transforming interactions of one domain into interactions of the destination domain, as described earlier in Chapter 6. Logically, the interdomain mapping function for any interdomain scenario can be considered to reside at the boundary between the domains or be split between them. There are two major approaches for the design of interdomain mappings depending on whether the transformation of interactions is done with or without some intermediate form.

- *Intermediate mapping:* Every element of interaction relevant to each domain is transformed, at the boundary of each domain, between the domain's internal form and some common, agreed form (via a mapper that may reside in a separate domain). It is important to mention that it is not necessary to establish one universal agreed form; more than one common form can be used in a single interdomain scenario with intermediate mapping. Each common form can then be optimized for a different purpose. In this case, the selection of which one is used can be done statically, for example through policies agreed between domain administrators, or dynamically between domains during runtime.

- *Direct mapping:* In this approach, every element of interaction relevant to each domain is transformed, at the boundary of each domain, directly between the internal form of one domain and the internal form of the other.

Another distinction among these approaches is regarding whether the components of the interdomain mapping functionality may be dispersed or are colocated.

8.3.3 Interdomain Object Referencing and Naming

For an entity (object, system, network, or subnetwork) to be accessible by its clients, some referencing mechanism is necessary. There are, of course, many referencing mechanisms possible that differ in their enforcement of features such as uniqueness, persistence, support for object migration, referencing schema, conventions and structure of valid references, and by the approach used to handle references across multiple domains. The term *object reference* usually presents some difficulties because of the overlap between the concepts of object identity, object reference, and object name.

Informally, we can say that the identity of an object is that which distinguishes it from all others, while a name or reference is a handle or token used to address an object [4]. Identity is difficult to define in a formal sense, but intuitively it is "something enduring about an object" [4]. ODP also focuses on identity as an essential feature that characterizes objects as "an object is distinct from any other object" [3], while names are a separate construct. For example, people do not change their identity if they change their name or telephone number. In CORBA, an object reference is a construct that reliably denotes a particular object. An object reference identifies the same object each time the reference is used in a request; and an object may be denoted by multiple, distinct references [2].

8.3.3.1 Global Referencing Spaces and Domain-Relative Referencing Spaces

A first, perhaps intuitive, approach to the problem of object referencing across multiple domains is to use a single, homogeneous global referencing space. This can only be achieved using a single referencing schema (i.e., a single referencing convention) and a single reference generation and registration authority. Then, each object reference is unique and the possibility that object references clash does not exist. This is the approach followed by ISO and by the Internet community regarding domain names, which must be registered in a central repository to ensure their uniqueness. However, this approach is very impractical for large referencing spaces composed of multiple domains not only due to scalability and maintenance issues but also because different domains often employ different referencing schemas and it is difficult to establish a single registration authority.

An alternative approach to global referencing spaces exploits the fact that objects in one domain do not really need to understand object references from the other domain, but just to associate the pair {Domain reference, Object reference} to some internal object reference. If we use the notation $\{D_0, R_0\}$ to denote an object reference R_0 from domain D_0, this is itself an object reference that can be used in multiple domains. We call this "domain-relative" referencing since it need not reflect the internal structure of object references within any domain. At an implementation level, associating an object reference with a domain is only important at an interdomain boundary. This approach is simple since the interdomain referencing function knows from which domain each object interaction comes, including any object references embedded in it.

Objects can also be addressed by means of names, where a name is generically defined as "a word or phrase that constitutes the distinctive designation of a person or thing" or "a word or symbol used in logic to designate an entity." Both object references and names refer to objects, but object references generally have no structure and are implemented in some way that facilitates effective machine processing; their main purpose, apart from denoting objects, is to "resolve predicates of sameness and equality" [4]. On the other hand, names are normally mnemonic and their structure and purpose is oriented toward human use. Object name spaces can also follow the global approach or the domain-relative approach.

In ODP and CORBA, a name-to-object association is called a *name binding*, which is always defined relative to a *naming context*. A naming context is an object that contains a set of name bindings in which each name is unique. Different names can be bound to an object in the same or different contexts at the same time. However, there is no requirement in CORBA that all objects must be named. To resolve a name is simply to determine the object associated with the name in a given context. To bind a name is to create a binding between a name and a reference in a given context. Thus, a name is always resolved relative to a context, and there is no need for absolute names [5]. For example, OSI naming (i.e., OSI object identifiers, OIDs) provides for both a global naming schema and a global name space (for all the constructs registered in the OSI OID registry).

There is a fundamental difference between a global name space and a global naming schema, concepts that mirror those of name and naming convention. In a global name space, any object that is a member of that name space has the same name anywhere. In a global naming schema, it is the naming convention that is unique everywhere that allows us to create name spaces that are global in the sense that any object anywhere can be named, but not necessarily by the same name everywhere.

In interdomain scenarios, the heterogeneity of referencing and naming mechanisms cannot usually be reduced to a single homogeneous scheme. When the goal is interoperability and sharing of resources in networked heterogeneous environments, common naming conventions are more important than global name spaces. Interdomain solutions also have to address the need for syntax-independent naming schemas, which is necessary for the federation of context-relative name spaces that use different naming syntaxes. Such solutions should use naming and name resolution in context-relative name spaces, that is, nested name spaces, hierarchical name spaces, and federated name spaces.

8.3.3.2 Interdomain Handling of Object References

In order to use context-relative name spaces in interdomain scenarios, models for referencing/naming entities in multiple domains and transformations of such references/names as they cross the domain boundaries are needed. Several basic schemes for accomplishing this exist; the following represent two possible approaches.

- *Reference encapsulation:* The handling of object references is accomplished by concatenating a domain identifier to the object reference each time a domain boundary is traversed. For example, if a reference R, originating in domain D_0, traversed domains D_1 and D_2, it could be identified in D_2 as reference $d_1.d_0.R$, where d_n is the domain identifier of D_n relative to D_{n+1}. Reference encapsulation adds domain information during the reference mapping process.

- *Reference translation:* The handling of object references is accomplished by providing encoded domain route information each time a domain boundary is traversed; the previous example would be as follows. R would be identified in D_2 as (d_1, x_1), and in D_1 as (d_0, x_0) where d_n is the domain identifier of D_n relative to D_{n+1} and x_n identifies the pair (d_{n-1}, x_{n-1}). Reference translation substitutes object references during the mapping process.

The following example illustrates these two approaches. We can imagine a galactic planet cartography system that catalogs planets (objects) in three planetary systems (domains): the solar system, Tau Bootis, and 16 Cygni B. The catalog is not maintained by a single central authority but distributed among astronomical societies in the three planetary systems. Each astronomical society uses a different classification scheme, but for simplicity we will assume that all use the same lexical convention, the ASCII alphabet. We now also imagine that astronomers in the solar system only have direct communication with colleagues in Tau Bootis and astronomers in 16 Cygni B can only communicate with peers in Tau Bootis. The solar system is cataloged as HD18009 in Tau Bootis, and Tau Bootis is cataloged as X_9_# in 16 Cygni B. What will an entry for planet Mars look like in each catalog? Table 8.1 illustrates the possibilities, depending on whether all three astronomical societies use reference encapsulation or reference translation (mixed scenarios are also possible). Table 8.1 also indicates the state information that has to be maintained at each point to ensure correct interdomain referencing. For example, HD18009:Mars <->4 means that when an object reference "Mars" comes from domain "HD18009" it is replaced internally by "4." Conversely, when an astronomer in Tau Bootis wants to obtain some information about object "4" in domain "HD18009," the astronomer will send a query about object "Mars" to some astronomer in the solar system. We note that there is a trade-off (probably more apparent if the example included 1,000 solar systems, several of them with nine or more planets) between complexity in the structure of references that are handed across domains and the state information that must be stored at each domain.

The problem of object identity could surface in this example if, for example, astronomers in 16 Cygni B received references to the object "Mars" from the solar system domain and from the Tau Bootis domain. Some procedure to check that two different references denote the same object would then be necessary (this is the case of *is_equal*( ) operators for object identity checking in object systems).

8.3.4 Handling Different Specification Languages

A very common issue in interdomain management arises from the fact that different domains can be specified using different specification languages. As

Table 8.1
Interdomain Object Referencing Example

Domain	Reference	State Information
Encapsulation		
Solar	Mars	
Tau Bootis	HD18009.Mars	HD18009
16 Cygni B	X_9_#.HD18009.Mars	X_9_#
Translation		
Solar	Mars	
Tau Bootis	(HD18009,4)	HD18009:Mars <-> 4
16 Cygni B	(X_9_#,d)	X_9_#:4 <-> d

we already mentioned, UML, CORBA IDL, and GDMO/ASN.1 are three commonly used nonequivalent specification languages. To enable a seamless solution for interdomain scenarios where domains have been specified in terms of different specification languages, it is necessary to find equivalent representations of objects in each domain using the specification language of the other domain. However, this is not always easy because the expressive power of two specification languages can be very different. When it is not possible to find equivalent constructs and mappings from one specification language to the other, then semantic information may be lost, and it would become necessary to use some additional mechanism to recover those semantics during run-time operation.

Interdomain scenarios with different specification languages usually also imply the use of different type systems. Two type systems are compatible if they support the same set of types, although some (or all) of the types may have different names in each domain. While compatibility between type systems on both sides of a domain boundary undoubtedly facilitates interdomain solutions, because a simple isomorphic mapping solves the interworking problem, it rarely occurs. Some of the technologies used for network and service management (e.g., CMIP, SNMP) have been primarily developed considering those specific applications, while others (e.g., CORBA) target multiple and diverse application realms. As a result of their differing objectives, the type systems for these technologies can differ significantly with respect to expressive power, naming scopes of identifiers, basic types and constructed types, subtyping, and even lexical conventions.

8.3.5 Interdomain Security

Security is a very complex issue and is a major concern in the design and deployment of computer systems and networks. A robust implementation of security policies and mechanisms in distributed object systems is a difficult task even when all the concerned applications and systems are under the same administrative authority but is clearly more difficult in interdomain scenarios. When dealing with different technology domains, some specific issues arise due to possible mismatches in the set of security capabilities that those technologies can offer at the interdomain boundary. In current telecommunications environments, as a first step, security is provided by deploying a closed network environment (i.e., one that does not provide unauthorized access to its resources, such as its network elements). In this book, we will not further address the issue of security in telecommunications networks.

8.4 CORBA/OSI Systems Management and CORBA/Internet Management Interdomain Interactions

For quite some time, the network management community has perceived the need to enable interoperation among multivendor management systems that may be based upon different technologies. The definition of interoperability between management systems at the level of physical system interfaces and protocols (e.g., CMIP, SNMP), as provided by Internet and TMN management standards, provides a first step in that direction. However, the software architecture of management systems was intentionally not addressed. As discussed in Chapter 2, further steps toward interoperability may be taken via enhanced specification techniques and usage of distributed computing infrastructures such as CORBA [2], which is an emergent standard on distributed object computing, defined by the OMG. While SNMP, CMISE, and CORBA have their strengths and weaknesses (as discussed to some extent in Chapter 2), they will undoubtedly coexist for quite some time. SNMP has a large embedded base in the general purpose computing market and the Internet; CMIP [6] is a powerful protocol with strong "on-the-wire" compatibility, mandated in the telecommunications arena by the ITU-T; and CORBA provides transparent support of heterogeneous platforms, operating systems, programming languages, and object distribution.

The main advantage of an interdomain approach in scenarios where these technologies coexist is that application designers would gain an effective environment in which to implement managers or agents and to integrate components from multiple vendors. For example, new managers could be built using CORBA components to manage existing CMIP or SNMP agents, reusing existing SNMP or CMIP MIBs. Thus, this approach enables objects in a CORBA domain to interact with remote GDMO [7] objects (or SNMP objects) as if they were in the CORBA domain; that is, the CORBA objects would see GDMO object interfaces as IDL-defined interfaces. The system designer would not need knowledge of GDMO, CMIP, or SNMP to accomplish this but only of CORBA and IDL. We see that this approach provides a cohesive path for the integration of these technologies, where each one may be used as most appropriate and existing management information models may be preserved for reuse.

8.4.1 Usage Overview

To enable interdomain interactions between CORBA and CMIP, or between CORBA and SNMP, it is necessary to map between their respective object models and provide mechanisms for handling conversion of interactions at the interdomain boundary. This mapping is split into two parts.

- The *specification translation* provides a mechanism for translating between IDL [2] and GDMO [7], or between IDL and SNMP MIB definition language (SNMPv2 MIB). Compilers that take, for example, a GDMO/ASN.1 specification and generate its equivalent IDL are called GDMO/IDL compilers.

- The *interaction translation* refers to the mechanisms needed to convert between the interactions in both domains, without either party being aware of the conversion. This dynamic mapping can be provided by a run-time gateway component.

A common scenario consists of a management system interface specified by means of a management information model; for example, a set of object classes that is defined in GDMO/ASN.1. This management information model is then translated via a specification translation algorithm into a set of IDL interfaces using a GDMO-IDL compiler. A manager, implemented as a set of CORBA objects, would manage objects supported by an OSI systems management agent as if they were CORBA objects (i.e., via the generated IDL interfaces plus some additional generic interfaces). These interfaces would be supported by a gateway supporting an interaction translation algorithm. The gateway terminates the standard CORBA generic inter-ORB protocol (GIOP) and dynamically translates CORBA invocations into CMIP messages based upon the original GDMO specification. The interaction translation mapping is bidirectional and translates CMIP messages originating from the OSI systems management agent into the appropriate CORBA requests and replies, as illustrated in Figure 8.2.

Conversely, if the manager is, for example, compliant with OSI systems management (as illustrated in Figure 8.3), it generates CMIP messages exactly as though the objects were supported by an OSI systems management agent. The CMIP protocol is terminated by the gateway, which dynamically translates the CMIP messages into GIOP requests on the CORBA objects.

The final case illustrates the use of CMIP as an environment-specific interoperability protocol (ESIOP). In this case, it allows both the manager and the agent to be implemented in the CORBA domain and yet to offer a standard external TMN interface. Neither party is aware of the implementation of the other, but the two back-to-back gateways ensure smooth operation of the overall system. While this is obviously less efficient than direct CORBA invocation, it does allow use of CMIP as an ESIOP, which could be useful in TMN environments.

Another scenario that may be covered by interdomain management is that of implementation of TMN architectures without using CMIP as the

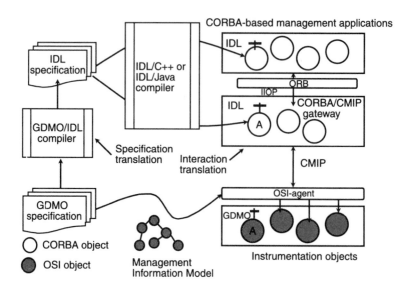

Figure 8.2 Specification translation and interaction translation.

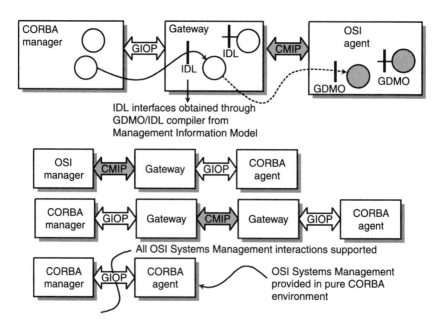

Figure 8.3 Usage scenarios in CORBA/OSI systems management interdomain interactions.

management protocol while preserving principles of OSI systems management, such as the manager/agent paradigm, operations with scoping and filtering, and notifications. This may occur when physically realizing X- or Q-interfaces in a pure CORBA environment when the management applications on both sides of the interface have been implemented using CORBA platforms. The protocol used at the interface is IIOP (versus CMIP), but the interface still complies with TMN principles.

The set of usage scenarios described in this section is illustrated in Figure 8.3 and is equally applicable to the CORBA/SNMP case. Using these scenarios, a CORBA manager may interact with OSI, SNMP, and CORBA agents.

8.4.2 Comparison of CORBA and OSI Systems Management

As we mentioned previously, an interdomain mapping function must provide for the transformation of allowed interactions in one domain to allowed interactions in the other domain. This can be a difficult task if both domains' models are not very similar to each other. As an example, we will examine a case involving CORBA/OSI systems management. While CORBA and OSI systems management have many similarities, they also have some important differences that complicate the specification and design of CORBA/OSI systems management interdomain solutions.

The CORBA and OSI systems management paradigms are based upon somewhat different models because each paradigm has been developed to meet specific business needs. In particular, OSI systems management was targeted toward network management and CORBA toward object-oriented distributed systems development. OSI systems management attempts to address the problem of interoperability at the syntactic and semantic levels but does not address code portability, while CORBA attempts to address both. However, the strategic objective of each model can be considered identical, that is, to promote interoperability between heterogeneous computing components. The models differ only in their approaches to specifying such interoperability and their choices of the boundary points at which interoperability is to exist.

A comparison of both paradigms in the computational viewpoint is provided next [8]. We introduce an appropriate computational description of OSI systems management, summarizing some similarities and differences among the two models, so the reader can understand some of the challenges faced in devising a CORBA/OSI systems management interdomain solution. The comparison among these paradigms encompasses the concepts of interoperability boundary points, specification techniques, object characteristics and taxonomy, object life cycle, object reference, naming, selection, and event notification.

- *Interoperability boundary points:* The choices of interoperability boundary points for OSI systems management and CORBA reflects their different concerns; OSI systems management is concerned with communications between management systems, while CORBA is concerned with distributed object systems specification and development. In OSI systems management, an agent that is resident in the managed system serves as an intermediary between the managing system and the managed objects. In CORBA, there is no intermediate agent; each object interface (as expressed in IDL) can be considered an interoperability boundary point.

- *Specification techniques:* The OSI systems management approach uses separate specifications for attributes, operations, and events that are bound into packages, qualified either as mandatory or conditional, which are subsequently bound into managed object classes. We note that conditional packages are a form of *type late binding.* While they meet OSI systems management's need for controlled variation in managed-object class specifications, they cause ambiguity for designers when used too liberally. In this model, the union of the object-class identification and the identification of the packages present serves as the indicator of a particular managed object's characteristics. Conditional packages complicate the type hierarchy for designers because the conditions for their presence are expressed in terms of natural language that cannot be formally evaluated. The OMG CORBA specification technique is more compact than the type hierarchy and does not allow type late binding. Specifications of attributes and operations are combined into interfaces, which may be further combined into modules, and reuse is achieved by inheritance (possibly multiple) of interface definitions.

- *Object characteristics:* With the exception of built-in operations to support manipulation of multiple attributes with a single invocation (allowed in OSI systems management but not in CORBA), differences between the models are very few. Both models enforce strict object encapsulation and provide distinct type systems for objects and attributes. In both cases, attributes cannot exist outside of objects, and built-in operations are provided for the manipulation of single attributes.

- *Object taxonomy:* The object taxonomies used for OSI systems management and CORBA are quite similar, with both making use of multiple-inheritance principles. The CORBA model supports the notion of substitutability by means of subtyping, and OSI systems management supports a similar notion by means of a limited subtyping capability termed compatibility.

- *Object life cycle:* OSI systems management provides built-in operations within the CMIP protocol for agents to create (M-CREATE) and delete (M-DELETE) object instances. Every object must be named upon creation, whereby the manager may either specify a name for the object or rely upon the agent to supply the object name. There is no standard defined for moving or copying objects between systems. CORBA does not provide built-in operations for instantiation of object interfaces; it uses a collaboration pattern based on factory objects (i.e., objects that create other objects), which allow clients to create specific object interfaces. On the other hand, the CORBA life cycle service [5] specifies standard interfaces for the deletion of objects and for copy/move of object implementations.

- *Object reference, naming, and selection:* Whereas CORBA provides both object references and names, in OSI systems management there is no specific concept of object reference other than the object distinguished name. CORBA object references are opaque types (their internal structure and representation are not visible to their users) of local significance only (may not be globally unique) and any object may have more than one reference.

Every managed object in OSI systems management must have a single distinguished name, composed of a sequence of relative distinguished names (RDNs); this sequence corresponds to a name scope hierarchy of managed object instance containment. Each RDN is a single attribute/value pair, assigned at the time of object creation, and may not be changed during the lifetime of the object. In the CORBA model, objects may have an arbitrary number of names (or no name at all). Names are distinct from objects and object references. The CORBA naming service [5] enables the use of context-relative, syntax-independent naming schemes.

Object selection in OSI systems management can be direct; a single object is selected by its name as the target of some operation, or multiple (zero, one, or more) objects are selected in a single construct as targets of some operation. Multiple selection is provided through scoping and filtering, where the scope may be the whole, or some subtree, of the managed object containment tree. CMIP provides built-in support for invocations with scoping and filtering, while CORBA provides for direct object selection but not a single mechanism for multiple selection through scoping and filtering. Several separate mechanisms may be used, such as the trader service [5], relationship service [5], or traversal of naming graphs through the naming service [5]. As we will see, the provision of scoping and filtering in CORBA, according to OSI systems management, is a fundamental requirement for the design of CORBA/OSI systems management interdomain solutions. We also note that as a consequence of differences in life cycle, object references, and naming, location transparency is not supported in OSI systems management but is in CORBA.

- *Event notification:* In OSI systems management events are forms of object behavior, with an explicit and distinct specification construct, reflected in a notifications template. Events are unconnected with object operations. There is a separate event report management function that determines which event notifications are forwarded to designated destinations and which are filtered out. In the CORBA model, the notion of events is accommodated into the event and notification services [5]. Objects that generate events, or objects interested in the reception of events, subscribe to the event or notification service (this may be done by invoking administrative operations on CORBA objects that are part of such services). Objects then generate events by invoking operations on the event or notification service. Distribution of events can be performed via "push" or "pull" models [5]. The push model allows a supplier of events to initiate the transfer of the event data to consumers. The pull model allows a consumer of events to request the event data from a supplier. In the push model, the supplier takes the initiative; in the pull model, the consumer takes the initiative. Event filtering is made possible using the notification service.

8.5 Joint Interdomain Management Solution to Interdomain Interactions

The Joint Interdomain Management (JIDM)[1] working group, jointly sponsored by the Open Group[2] and the Network Management Forum (now known as

1. Joint Interdomain Management working group, jointly sponsored by the NMF and the Open Group to develop specifications that facilitate interworking between CMIP-, SNMP-, and CORBA-based management systems. More details may be found at URL http://www.jidm.org.

2. The Open Group is an industry forum that was established to answer fundamental questions on IT (information technology) by aiding in the development and implementation of a secure and reliable IT infrastructure. It concentrates on protecting investment in heritage systems while helping organizations evolve their computing architectures to the requirements for a global information infrastructure. More details may be found at http://www.opengroup.org.

the TeleManagement Forum), was chartered to work on a solution to address CORBA/OSI systems management and CORBA/Internet management interdomain interactions. Previously, the Network Management Forum ISO-Internet Management Coexistence (IIMC) group addressed SNMP/CMIP interdomain scenarios [9]. These relationships are illustrated in Figure 8.4.

The CORBA/OSI systems management solution proposed by the JIDM was adopted by the OMG and provides both specification translation and interaction translation solutions to enable interdomain management in multivendor environments. Both solutions are complete, reuse existing CORBA services, and do not mandate any particular CORBA implementation. In this section, we focus upon the approach taken toward addressing CORBA/OSI systems management interactions. The interested reader can find detailed information regarding the approach to CORBA/SNMP interdomain interactions in [8, 10].

8.5.1 Specification and Interaction Translations

Specification translation [8] covers the static process by which specifications are translated from one specification language to another, while interaction translation covers the run-time process by which interactions from one domain are mapped onto interactions from the other domain. We note that a simple one-to-one mapping is not always possible, and thus a single interaction in

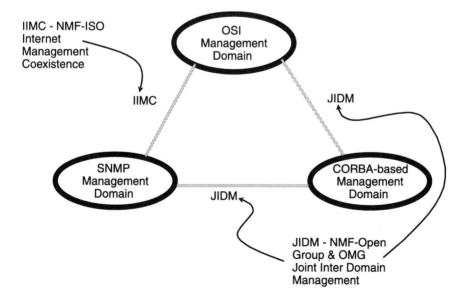

Figure 8.4 CORBA, OSI, and Internet interdomain management.

one domain must sometimes be transformed into multiple interactions in the other domain.

8.5.1.1   Specification Translation

We will first consider ASN.1 [11], which has a more complex type system than IDL [2]; as a consequence, some information is necessarily lost in the translation process from ASN.1 to IDL. For example, ASN.1 tag values and subrange types do not have a representation in IDL, as follows.

When a size constraint is used for either SET OF and SEQUENCE OF constructions in ASN.1, it is mapped to IDL as a bounded **sequence**, where the upper bound of the IDL sequence is determined by the upper bound on the permitted integer values specified in the constraint. In addition, a constant giving the permitted values list is provided.

```
T1  ::= SEQUENCE SIZE(0..16) OF LogRecord
T2  ::= SEQUENCE SIZE(16|32|64) OF INTEGER
```

maps to IDL as follows:

```
typedef sequence<LogRecordType, 16> T1Type;
// T1Type SIZE(0..16)
typedef sequence<ASN1_Integer, 64> T2Type;
// T2Type SIZE(16|32|64)
```

To ensure validity of ranges during run-time operation, the information lost during specification translation is captured in run-time code. In this case, the interaction translation mechanism will have dynamic access to size information as defined by the IDL comments and will ensure that the ranges for those types are restricted to the defined values.

Several issues have to be resolved in the specification translation process. Those for the GDMO/ASN.1 to IDL translation are elaborated as follows.

- *Lexical translation:* ASN.1 names are case-sensitive but IDL identifiers are not. In addition, ASN.1 has different name spaces for type-references, identifiers, and value-references, whereas IDL has a single name space for all those elements. Both have scoped naming spaces, but the naming scopes do not map directly; for example, enumerated types create a new naming scope in ASN.1 but not in IDL.

- *Mapping of constants:* IDL only permits constants of primitive types (integer, boolean, floating-point, character, and string types). As such, it cannot adequately represent constant values of more complex ASN.1

types. Some of these ASN.1 constants are often used to define default values; therefore, their equivalent IDL representation must provide simple and safe access to those constants.

- *Mapping of GDMO templates:* An IDL interface is defined for every GDMO managed object class template, which supports the operations exported by members of the corresponding managed object type. In addition, a module is defined that contains interfaces supporting notifications. For each action, an IDL operation is declared. If the action has a WITH REPLY SYNTAX clause, then the necessary IDL to support the handling of multiple replies during run time is generated. GDMO attributes are not mapped directly to IDL attributes but to IDL operations. The reason for this lies in the different semantics associated with attribute access permissions in GDMO and in IDL. Each attribute in a GDMO package has a property list that identifies the access permissions for that attribute. These may be GET, REPLACE, GET-REPLACE, ADD, REMOVE, ADD-REMOVE, or REPLACE-WITH-DEFAULT. In addition, CMIP may raise a number of different errors (possibly with parameters given in SPECIFIC-ERROR parameters) such as `processingError` or `illegalValue`. These are mapped to IDL as user exceptions so that the additional data may be carried, which is the main reason why GDMO attributes are not mapped directly to IDL attributes. User exceptions cannot be raised as a result of access operations on IDL attributes, but they can be raised as a result of invoking regular operations in IDL.

8.5.1.2 Interaction Translation

Interaction translation [10] covers the run-time process by which interactions from one domain are mapped onto interactions from the other domain. As mentioned earlier, a single interaction in one domain must sometimes be transformed into multiple interactions in the other domain. For example, if a scoped and filtered CMIP GET request is received, an OSI/CORBA gateway would have to identify the set of CORBA objects matching the filter within the scope and invoke the appropriate operation on each of those objects. The results would then be collated and formatted into one or more CMIP messages in the reply.

Thus, a CORBA/CMIP gateway must be able to:

- Translate any incoming CMIP SET/GET/ACTION request into one or more invocations to methods supported by some CORBA object(s) in the CORBA domain;

- Receive any event generated by a CORBA object from the CORBA domain and translate it into an EVENT-REPORT request to be forwarded to remote OSI systems in the OSI domain that had registered their interest in receiving events;

- Receive incoming CORBA invocations from objects in the CORBA domain and forward them as CMIP SET/GET/ACTION requests to some OSI agent(s) in the OSI domain;

- Receive CMIP EVENT-REPORT requests from the OSI domain and forward them as CORBA events to interested parties in the CORBA domain;

- Receive any incoming CMIP CREATE/DELETE request from the OSI domain and translate it into an invocation to a method being supported by an object (e.g., a factory object) in the CORBA domain;

- Receive CORBA invocations for creating/deleting objects in the CORBA domain and forward a CMIP CREATE/DELETE request to some OSI agent in the OSI domain.

These transformations must preserve all the semantics required by OSI systems management regarding how objects are created and deleted, how they are named and can be accessed, and how they generate and forward event notifications.

8.5.2 Reference Model

The JIDM solution defines a basic reference model (illustrated in Figure 8.5) that extends the manager/agent model to provide a common management framework valid for CORBA, OSI systems management, or Internet management scenarios. A distributed object management system is viewed as composed of two kinds of objects: manager objects and managed objects. Manager objects are those that are responsible for one or more management activities by issuing management operations and receiving event notifications. Managed objects perform management operations issued by manager entities on the underlying resources and emit event notifications whenever some specific circumstances occur.

These objects are grouped into domains, which form the unit of accessibility; manager domains group manager objects, while managed domains group managed objects. Managed domains are sometimes referred to as agents or managed object domains, while manager domains are sometimes referred to as manager applications or simply managers. Domains are identified by means of titles, which may have different forms for the different environments. For

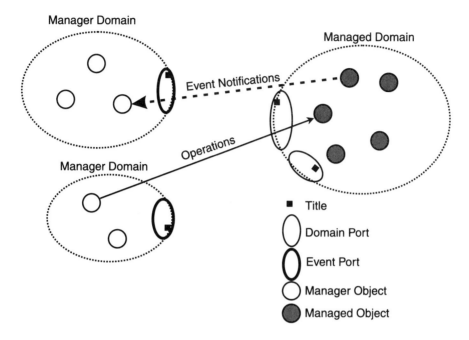

Figure 8.5 The JIDM reference model.

example, OSI environments use an application entity title (an identifier for an OSI application entity, which can be loosely defined as an OSI application layer software application or component), SNMP environments use an IP address or hostname, and pure CORBA environments use an interoperable object reference (IOR) or CORBA name. Each domain may have more than one title associated with it, but a given title uniquely identifies one domain.

Whenever a domain needs to interact with another domain, it must first gain access to the other domain. This access is always granted though a specific port to the domain. Two types of ports are identified in the JIDM reference model:

- *Domain port:* Used to grant and control access to a domain to create and invoke operations on managed objects;
- *Event port:* Used to grant access to a domain to forward event notifications to manager objects within that domain.

8.5.3 Interface Taxonomy

The JIDM solution provides a set of IDL interfaces that can be used to implement and deploy CORBA/CMIP or CORBA/SNMP interdomain solutions and also to deploy OSI systems management or Internet management solutions in pure CORBA environments. To cover all the different possible scenarios with a minimum of redundancy, JIDM defines a basic set of IDL interfaces, referred to as JIDM facilities, common to any management reference model (CORBA, OSI systems management, or Internet management). JIDM facilities can be extended to provide additional functionality that is specific to one of the management paradigms but still common to any particular management information model. These IDL interfaces are referred to as OSI management facilities and SNMP facilities, and their taxonomy is provided in Figure 8.6.

In summary, the IDL interfaces specified by JIDM can be categorized as:

- *Generic interfaces, management paradigm independent:* These facilities provide a generic framework with mechanisms to access a managed domain independently of the management reference model being used (OSI systems management, Internet management, or CORBA). These generic IDL interfaces are referred to as JIDM facilities and are collected in the JIDM module. Proper collaboration among these interfaces and some others from standard CORBA services (naming, life cycle, event, notifications) provide management applications with facilities for boot-

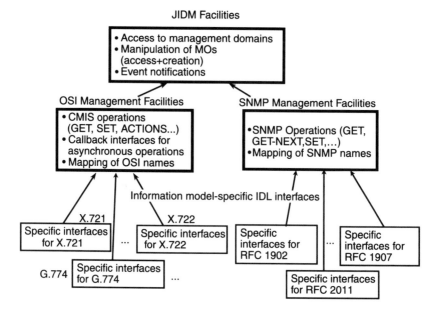

Figure 8.6 Taxonomy of JIDM IDL interfaces.

strapping, access control, creation and deletion of managed objects, localization of managed objects, invocation of operations on managed objects, and distribution of event notifications.

- *Generic interfaces, management paradigm dependent:* These facilities provide additional generic mechanisms that are specific to one management paradigm, such as OSI systems management or Internet management.

  - OSI management facilities provide a CORBA view of the OSI systems management reference model. This set of IDL interfaces extends the generic JIDM facilities to support all possible CMIS interactions in CORBA and concepts specific to OSI systems management. Such concepts include the ability to invoke operations using scoping and filtering and to create, delete, and name objects according to OSI systems management principles, both in pure CORBA environments as well as within interworking environments.

  - SNMP management facilities provide a CORBA view of the Internet management reference model. This set of facilities also extends the generic JIDM facilities to support all SNMP interactions in CORBA and Internet management specific concepts.

- *Specific interfaces, information and management model dependent:* These IDL interfaces provide functionality that is specific to a given management information model conforming to a certain management reference model; these interfaces reuse and extend OSI management or SNMP management facilities in an information model-specific way. If the specific management information model is specified in GDMO/ASN.1 or in SNMP SMI, the translation algorithms defined in the specification translation part of JIDM may automatically generate the equivalent CORBA IDL model. We note that it is also possible to specify an information model directly using IDL and yet reuse the OSI management or SNMP management facilities.

8.5.4 Coarse- and Fine-Grained Approaches to OSI Systems Management in CORBA

OSI management facilities provide IDL interfaces and operations that satisfy both coarse-grained and fine-grained approaches to the problem of providing OSI systems management interactions in pure CORBA scenarios and in interworking scenarios.

The coarse-grained approach is so called because it represents all the managed objects in a managed domain by a single CORBA object: the *OSIMgmt::ProxyAgent.* This interface exports CMIS-like operations—such as *cmis_create( )*, *cmis_delete( )*, *cmis_create_sync( )*, *cmis_set( )*, *cmis_get( )*, *cmis_action( )*—that can be used by manager objects to provide OSI systems management interactions. This approach is truly minimalist, but it obviously requires detailed knowledge of CMIP to properly build applications.

On the other hand, the fine-grained approach is a more natural approach because it models each managed object by a single CORBA object (or interface to be more precise). The design of management applications using this approach benefits from all the features of object-oriented modeling and design, such as encapsulation, separation of concerns, and generalization and specialization. This approach does not require specific knowledge of CMIP to engineer OSI systems management compliant applications, and it provides transparency with regard to whether the applications are deployed in pure CORBA environments or within interworking scenarios. Moreover, if the management information model is specified in GDMO/ASN.1, then the individual IDL interfaces that model managed objects can be generated automatically using a GDMO-IDL compiler. We note that this approach does not imply an inherent scalability problem because newer implementations of JIDM facilities and CORBA/CMIP gateways avoid the problems associated with creating and maintaining large numbers of CORBA objects in a system.

8.5.5 The OSI Systems Management Managed Object Interface

In OSI systems management all managed objects inherit from *X721::top*; OSI management facilities introduce an inheritance hierarchy to respect this requirement and a base interface that collects all the expected semantics that are specific to OSI systems management (e.g., scoping and filtering, operations on multiple attributes). The standard X.721::top interface inherits from OSIM-gmt::ManagedObject interface. As a consequence, all management interfaces generated by a GDMO to IDL compiler inherit (indirectly) from the *OSIM-gmt::ManagedObject* interface. This inheritance tree is shown in Figure 8.7.

8.5.6 CORBA/CMIP Gateways

We now briefly explore one particular scenario of CORBA/OSI systems management interworking, specifically, CORBA manager objects interworking with a CMIS agent through a CORBA/CMIP gateway. Using JIDM facilities, CORBA manager applications create and invoke operations on managed objects in the same way they create and invoke operations on ordinary CORBA objects located in the same CORBA domain. Analogously, they receive events supplied by managed objects as if they were ordinary CORBA objects supplying events to an event channel located in the CORBA domain. How this actually happens

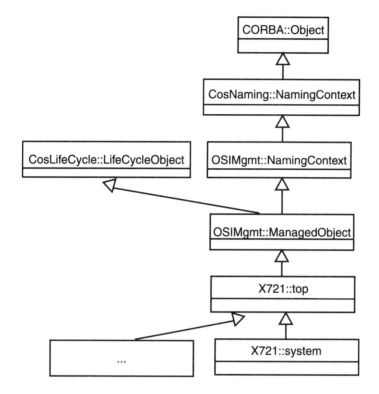

Figure 8.7 OSI systems management inheritance tree.

is transparent to the CORBA manager program. This allows the provision of all the possible interactions in OSI systems management in pure CORBA environments, or in CORBA/CMIP interworking scenarios, in a way that is totally transparent to the CORBA manager objects.

CORBA/CMIP gateways must be used by any CORBA manager application needing to interoperate with managed object domains that are not directly accessible via CORBA but are accessible via CMIP. A CORBA/CMIP gateway has several CORBA objects associated with it (as illustrated in Figure 8.8):

- A *JIDM::ProxyAgentFinder* object for establishing connections to OSI managed object domains accessible via CMIP through the gateway;
- One or several *JIDM::EventPort* objects for receiving notification of events from members of OSI managed object domains accessible via CMIP through the gateway.

The *JIDM::ProxyAgentFinder* object is created during start-up of the CORBA/CMIP gateway. *JIDM::EventPort* objects at the gateway may be

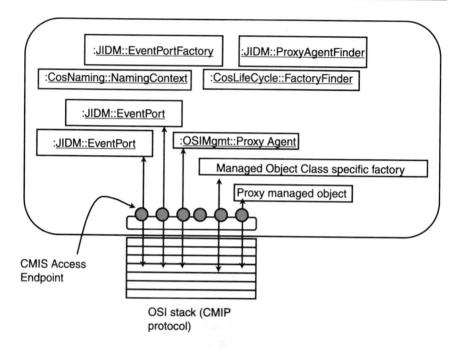

Figure 8.8 CORBA/CMIP gateways.

created during or after start-up. The latter case requires the existence of a *JIDM::EventPortFactory* object at the gateway. As a result of establishing a connection through a CORBA/CMIP gateway, an *OSIMgmt::ProxyAgent* object is created at the gateway. *OSIMgmt::ProxyAgent* objects created this way are responsible for:

- Creating a *CosLifeCycle::FactoryFinder* object that enables creation of CORBA factories that handle creation of managed objects at the domain;

- Creating a *CosNaming::NamingContext* object that enables creation of CORBA proxy managed objects for each member of the domain;

- Sending scoped operation requests.

How do JIDM objects and the CORBA/CMIP gateway interact to provide the expected behavior? In the following subsections, we briefly explain the interaction patterns used to provide four fundamental behaviors: getting access to managed object domains, creating managed objects, invoking operations on single managed objects, and receiving and distributing events.

8.5.6.1 Getting Access to Managed Object Domains

The following steps are followed when a CORBA manager object tries to get access to a managed object domain using a CORBA/CMIP gateway (as shown in Figure 8.9).

- The CORBA manager object invokes the *access_domain( )* operation exported by the *JIDM::ProxyAgentFinder* object located at the gateway. Information that unequivocally identifies the managed object domain to be accessed is passed in the invocation.

- As a result of invoking the *access_domain( )* operation, an *OSIMgmt::ProxyAgent* CORBA object is created at the gateway. This *OSIMgmt::ProxyAgent* object is bound to a CMIP communication endpoint (a CMIS access point). A connection is established with the managed object domain whose domain title was specified in the criteria passed as argument to the *access_domain( )* operation.

- A reference to the *OSIMgmt::ProxyAgent* object is returned to the CORBA manager object that requested access to the managed object domain being considered. That reference identifies the managed object domain in the CORBA space.

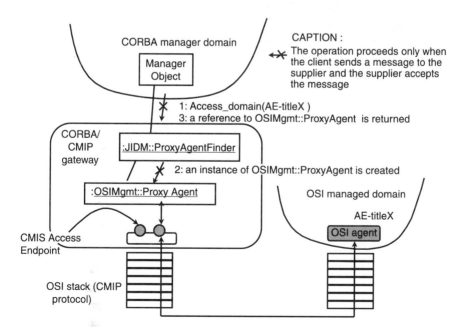

Figure 8.9 Getting access to a managed object domain.

8.5.6.2 Creating Managed Objects

The following steps are followed when a CORBA manager object creates a managed object (e.g., SSCC-X of class SSCC) at some domain that is accessible through a CORBA/CMIP gateway (as shown in Figure 8.10):

- The CORBA manager object invokes the *get_domain_factory_finder( )* operation exported by the *OSIMgmt::ProxyAgent* object.

- The CORBA manager object invokes the *find_factories( )* operation exported by the returned *CosLifeCycle::FactoryFinder* object, passing a valid key for the object to be created.

- The *CosLifeCycle::FactoryFinder* object finds references for appropriate managed object factories at the gateway. If there is no managed object factory matching the key, the *CosLifeCycle::FactoryFinder* object creates one. References to managed object factories are returned to the CORBA manager.

- The CORBA manager object invokes an operation (e.g., *create( )*) on the managed object factory using the CORBA object reference it obtained to request creation of the managed object in the OSI managed domain.

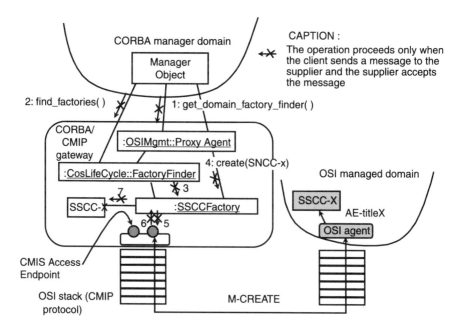

Figure 8.10 Creating managed objects.

- The CORBA request is received by the CORBA/CMIP gateway and is translated into an appropriate M-CREATE message. This M-CREATE message is sent through the association handled by the *OSIMgmt:: ProxyAgent*.
- When the response to the M-CREATE message is received by the gateway, the invoked operation returns with the appropriate result values.
- A CORBA proxy managed object (SSCC-X) is created at the gateway and a reference to it is returned to the CORBA manager object; this reference can then be used to denote the remote OSI object.

8.5.6.3 Invoking Operations on Individual Managed Objects

The following steps are followed when a CORBA manager invokes an operation on a managed object (e.g., SSCC-Y) that is accessible through a CORBA/ CMIP gateway (as illustrated in Figure 8.11). To make the example more interesting, we suppose that the manager object does not hold a reference to a CORBA object that acts as proxy of the remote OSI managed object but only has its name (e.g., "systemW.SSCC.c1").

- The CORBA manager object invokes the *get_domain_ naming_context( )* operation exported by the *OSIMgmt::ProxyAgent* object in the gateway.

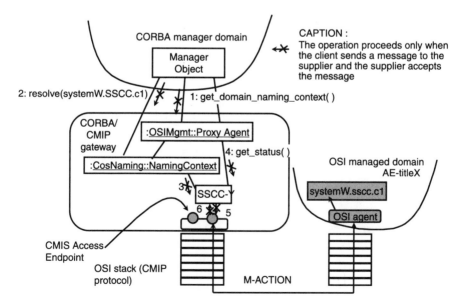

Figure 8.11 Invoking operations on individual managed objects.

- The CORBA manager object invokes the *resolve( )* operation exported by the returned *CosNaming::NamingContext* object, passing the name of the managed object upon which it wants to operate.

- The *CosNaming::NamingContext* object finds a reference to the CORBA object acting as the proxy of the managed object and returns it to the CORBA manager object that requested it.

- The CORBA manager object invokes an operation (e.g., *get_status( )*) on the managed object using the CORBA object reference to the corresponding proxy object.

- The CORBA request is received by the CORBA/CMIP gateway and translated into an appropriate CMIP message (M-ACTION). This CMIP message is sent through the association handled by the *OSIMgmt::ProxyAgent.*

- When the response to the CMIP message is received, the invoked operation returns with the appropriate result values.

8.5.6.4 Receiving and Distributing Events

Events originated at managed object domains are received through *JIDM::EventPort* objects at CORBA managers. A mechanism is implemented at the CORBA/CMIP gateway that allows event data received at a management connection endpoint to be forwarded to the appropriate *JIDM::EventPort* object. The following steps are followed when a CORBA manager object receives an event through a *JIDM::EventPort* at a CORBA/CMIP gateway (as illustrated in Figure 8.12):

- In the start up phase of the CORBA manager application, one or more CORBA objects (including intermediate EventChannel or NotficationChannel objects) register themselves in one or more of the existing *JIDM::EventPort* objects either as *CosEventComm::PushConsumers* or *CosEventComm::PullConsumers.*

- An M-EVENT-REPORT CMIP message containing notification of an event from a managed object is received by the CORBA/CMIP gateway through some association. This association is bound to a specific title and has a *JIDM::EventPort* object associated with it that receives the event data carried in the CMIP message.

- The appropriate response is sent by the CORBA/CMIP gateway back to the managed domain that reported the event, confirming that the event was received at the manager domain.

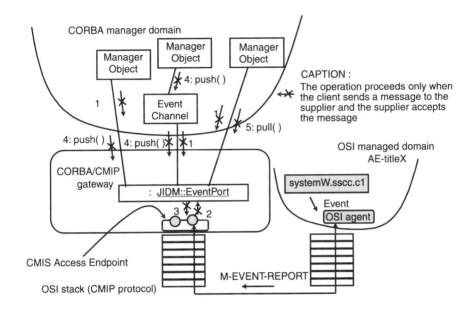

Figure 8.12 Receiving and distributing events.

- The *JIDM::EventPort* object invokes the *push( )* operation exported by all *CosEventComm::PushConsumers* objects connected to it.

- The *JIDM::EventPort* maintains the event until all *CosEvent-Comm::PullConsumers* objects connected to the port pull the event.

8.6 Summary

The feasibility of designing and deploying interdomain solutions is of special importance due to the heterogeneity and distributed nature of today's networks and services. In this chapter, we explored some of the issues associated with interdomain management and, in particular, interdomain interactions. We also briefly explored the particular problem of interworking between CORBA and OSI systems management domains, and one solution, JIDM, that has reached standards status. The interdomain management results described herein may be easily extended for application to any other kind of interdomain relationship. For example, the current evolution of intelligent networks (IN) toward the support of CORBA infrastructures (as a result of TINA-C work efforts) can employ the JIDM specification translation. This opens the door for convergence between IN and TMN.

References

[1]   Scott, D. S., and C. Strachey, "Towards a Mathematical Semantics for Computer Languages," *Proc. Symposium on Computers and Automata,* Microwave Research Institute Symposia Series, Vol. 21, 1971.

[2]   The Object Management Group, "Common Object Request Broker: Architecture and Specification," revision 2.2, Feb. 1998.

[3]   ISO/IEC 10746-2: ITU-T Recommendation X.902 "Information Technology, Open Distributed Processing, Reference Model: Foundations," 1995.

[4]   Kent, W., "A Rigorous Model of Object Reference, Identity, and Existence," *Journal of Object-Oriented Programming,* Vol. 4, No. 3, June 1991.

[5]   "CORBA Services: Common Object Services Specification," Nov. 1997.

[6]   ITU-T Recommendation X.711 | ISO/IEC 9596-1:1991, "Information Technology, Open Systems Interconnection, Common Management Information Protocol, Part 1: Specification."

[7]   ITU-T Recommendation X.722 | ISO/IEC 10165-4:1992, "Information Technology, Open Systems Interconnection, Structure of Management Information, Part 4: Guidelines for the Definition of Managed Objects."

[8]   Joint X/Open-NMF Inter-Domain Management (JIDM) Task Force Specification Translation. X/Open Preliminary Specification, May 1997.

[9]   "CMIP/SNMP Interworking; TeleManagement Forum Component Set CS341," Issue 1.0, Dec. 1995.

[10] JIDM Interaction Translation. Final Submission to OMG's CORBA/TMN Interworking RFP, edition 4.3, Oct. 1998; OMG Document Number: telecom/98-10-10.

[11] ITU-T Recommendation X.208 | ISO/IEC 8824:1990, "Information Technology, Open Systems Interconnection, Specification of Abstract Syntax Notation One (ASN.1)."

9

Multitechnology Application Example

9.1 Introduction

In this chapter, we provide a concrete example that illustrates application of all of the concepts and models we defined in the preceding chapters. Our intention is to illustrate the usage and interrelationships among these concepts and models via a complete example that involves specification of a particular service and illustrates how this specification translates into associated network and equipment specifications. While the example will not be elaborated in exhaustive detail, it will be sufficient to enable the reader to understand how a complete set of specifications may be derived.

9.2 Scenario Description

In this section, we provide a high-level enterprise description of a particular service supported over a transport network involving a customer who has asymmetric digital subscriber line (ADSL) access at home supporting his or her telephone, multimedia, and television usage. This customer is interested in having a video on demand (VoD) service that may be accessed via an Internet session on his or her PC. When the customer initiates an Internet session to request a particular video, the video distributor receiving the request selects a particular video server, which in turn requests an ATM connection carried over an SDH core network to carry the video stream and thus provide the service. We will illustrate the business concerns relevant to this scenario using the TINA-C business process model discussed in Chapter 6. After providing a summary of the customer and VoD service provider perspectives of the service, we will focus upon connection management in a multimanagement domain and multitechnology environment.

As explained in Chapter 6, the relevant business concerns may be reflected in the establishment of commercial relationships between roles that are played by actors, with the associated models expressed using ODP-based enterprise viewpoints. Thus, service expectations are described within a set of contracts (referred to as service level agreements, or SLAs in the TeleManagement Forum), each instance of which enumerates associated roles, policy statements, and enterprise actions. In this chapter, we use the existing frameworks provided by TINA-C and specified in more detail within the ITU-T Global Information Infrastructure Recommendations [1–3] for describing business roles to assist in the development of our enterprise models. The associated enterprise viewpoint specifications will be documented via UML diagrams as described in Chapter 7. This also assists us, via provision of appropriate transformation tools, in relating enterprise specifications to their associated information and computational viewpoint specifications according to the rules expressed in both Chapters 4 and 7.

9.2.1 Customer Perspective

Let us consider a residential client customer, Mr. Smith, who subscribes to a VoD service provided by a video distributor. Mr. Smith orders movies using his PC and then views them on his TV. In addition to his subscription to the video service, Mr. Smith also subscribes to an Internet access service (provided by the Internet access provider, IAP_I) and an access network service (provided by the operator, telco Y). The contract between Mr. Smith and the video distributor company stipulates that a fixed charge of $5 will be incurred per movie request. As part of the movie-ordering phase, Mr. Smith has to identify himself by providing some authentication of his identity. At the end of each month, he receives a bill from the video distributor company. However, as a service guarantee, the video distributor company stipulates in its contract with Mr. Smith that any performance impairment resulting in an interruption of the film display persisting for more than one minute will result in the movie being provided free of charge.

According to the methodology of Rec. G.851 described in Chapter 4, the aforementioned contract-related points are considered to be obligations in the contract between Mr. Smith and the video distributor company and can be expressed as follows:

Obligation: price per viewing
Each movie has a $5 fee.

Obligation: identification/authentication
The client needs to provide identification and authentication to gain access to the service.

Obligation: billing schedule
A bill will be sent to the client by the 25th of each month.

Obligation: service guarantee
There will be no charge for the service for any failures persisting for more than one minute.

The services contracted by Mr. Smith are summarized in Figure 9.1; clearly, Mr. Smith can also have other service subscriptions outside of the scope of this example.

9.2.2 Video Service Provider Perspective

The video distributor does not serve as the repository of the movies; rather, these movies are available on several video servers for which the video distributor

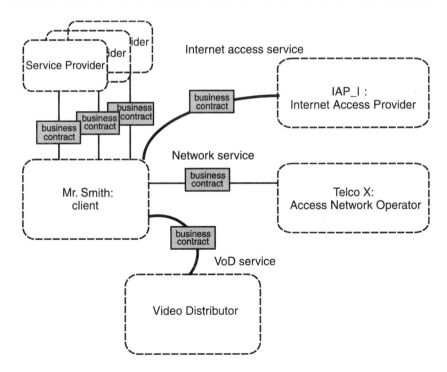

Figure 9.1 Customer perspective of VoD service.

company acts as a client (as illustrated in Figure 9.2). Mr. Smith does not have, and does not need to have, any knowledge of the existence of these video servers. Thus, the video distributor company plays a retailer role in the TINA-C business model, while each video server acts as a third party provider. We additionally note that the video distributor company will have many contracts with residential customers. In our scenario, the video distributor company provides a bill at the end of each month to each of its customers and has to pay the video servers for providing the distributed movies. This discussion illustrates the concepts of contract type and contract instance explained in Chapter 4. Specifically, the company has a contract type for video service, but for each customer a contract instance is elaborated that may include or exclude some conditional features that are part of the contract type. For example, such a feature could be user mobility, which would be expressed in the contract as a permission.

Permission: user mobility
In addition to the main location, the subscriber may have access to the VoD service in a secondary location, providing that VoD sessions cannot be established in parallel at the two locations.

While in our particular scenario Mr. Smith controls the VoD service using his PC, in reality the customer may control their VoD service from a television (or even a telephone), as requiring a customer to maintain separate subscriptions solely for the purposes of service usage and service usage control

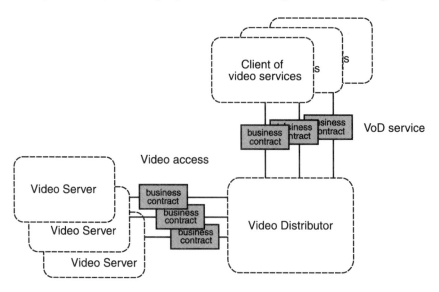

Figure 9.2 VoD service from the provider perspective.

would not be good commercial practice. However, the physical separation of the PC from the television in this example more easily illustrates the logical separation of access and usage sessions and assists in explaining the principles used within the chapter. Of course, we recognize that engineering decisions (i.e., ODP engineering viewpoint) determine the selection of network protocols and whether access and usage take place from the same terminal. As we are aware, these decisions have little to do with the enterprise specification of the VoD service.

In this scenario, the video distributor company has a contract with an Internet access provider IAP_J, as illustrated in Figure 9.3, which obviously may be different from the Internet access provider directly contracted by Mr. Smith (IAP_I). We assume that IAP_I and IAP_J can cooperate, which allows the video distributor company and Mr. Smith to exchange VoD service control information over the Internet.

Within this scenario, control of the VoD service by the video distributor company involves establishing a stream connection between the selected video server and Mr. Smith's television and control of this stream (e.g., stopping the

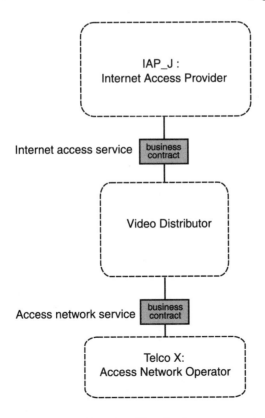

Figure 9.3 Communication aspects from the video distributor perspective.

billing if a service usage-related failure is detected). These activities imply that the video distributor company is a client of control and management services offered by the access and core network telecommunications operators for all of the video distributor company's VoD service customers and the associated video server companies it may need to access. Here, the video distributor company is not directly responsible for the global coordination of the end-to-end telecommunications connection; it delegates this responsibility to the network provider, telco Y.

Thus, we see that in order to control and manage the VoD service, the video distribution company must be a client of telecommunications services and needs to be able to access the telecommunications network. As such, in our example, the video distributor company is a client of an access network operator, telco Y. Obviously, the video servers also need to be connected to access networks operated by telecommunication companies. To simplify the example, we show only one video server, which has a business contract with the same access network as the video distributor company. At the end of each month, the video distributor will be charged by telco Y for the network resources it has used to support the VoD service. Thus, the $5 charge per movie includes the fees for video and network resource usage as well as service packaging.

9.2.3 Telecommunications Operator Perspective

Following the terminology used within the ITU-T GII Recommendations [1–3], we classify telecommunication domains into *access* and *core,* which are related to each other via business contracts. As illustrated within Figure 9.4, these business contracts support the provision of transport capabilities for stream interworking and the provision of control and management capabilities for control and management interactions spanning the access and core network domain boundary. As demonstrated in Chapter 4, such contracts eventually lead to agreements concerning engineering interfaces. Additionally, other engineering interfaces exist within individual domains, which may or may not be derived from an explicit contract. For example, as will be described in Section 9.3, the core network is topologically comprised of two interconnected SDH rings, where it is clear that a contract must have been established between the telecommunications equipment vendors that delivered the SDH rings and the service provider, telco Y, that purchased them. Such contracts involve more than just equipment usage; they inherently include associated constraints, some of which arise from conformance to standards.

9.2.3.1 Access Network Operator

The access network operator establishes business contracts with customers to enable access networks to be connected with customer premises equipment or

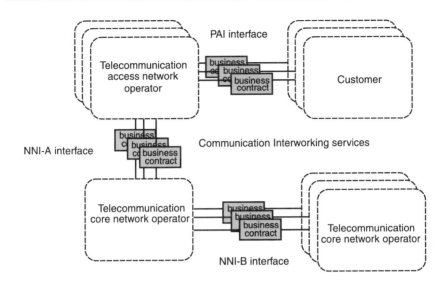

Figure 9.4 Communication aspects from operator perspective.

networks at *Premise-Attachment-Interfaces* (*PAI*) [1]. The access network operator similarly establishes business contracts with core network operators to enable
the access network to be interconnected with the core network at *Network-to-
Network-Interface type A* (*NNI-A*) [1]. Within the scenario we are considering,
two access network interconnections are established: one interconnection is
within telco X's domain, which enables access (via ADSL and ATM) to Mr.
Smith's customer premises equipment, and the second is within telco Y's
domain, which provides access to the video distributor and the video server
premises (via SDH and ATM). Telco Y is responsible for global coordination
of the transport capabilities needed to support the VoD service and will therefore
charge the video distributor for the provided network services. Telco Y assesses
the total bill by processing the charging information received from its access
and core subdomains and telco X's domain.

9.2.3.2 Core Network Operator

In addition to the NNI-A interfaces described previously, the core network
operator may also establish business contracts with other core network operators
at *Network-to-Network-Interface type B* (*NNI-B*) [1]; however, they do not
establish business contracts directly with end customers. For simplicity, we
consider only one core network operated by telco Y to provide connectivity
between the two access networks, operated by telco X and telco Y, respectively
(this will be illustrated in Figure 9.5 in the following section).

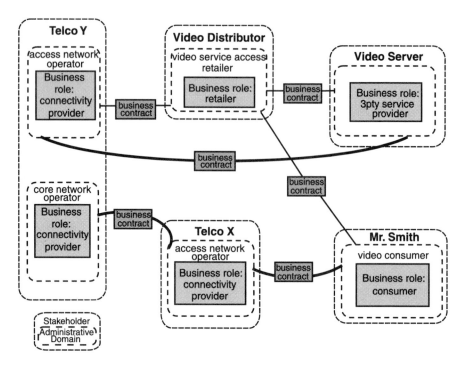

Figure 9.5 TINA-C business model applied to the VoD service scenario.

9.2.4   Business Model of the Scenario

We expect some common behavior among the various types of business contracts enumerated in the previous sections. For example, it is clear that the nature of contracts with telecommunications operators is all very similar, in that they all deal with the provision of connectivity and associated transport capabilities. In the same way, we should be able to assume that a contract with an Internet access provider is just a specialization of a more general service access contract. We make this assumption by adopting a more generalized approach, such as that offered by the TINA-C business model, which acts as a framework for service access and service usage, as discussed in Chapter 6. Using such a model allows us to further reduce complexity by developing specific roles within each domain and for each business contract, which are linked to the generic retailer, connectivity provider, and third-party service provider roles already defined in TINA-C. The result of applying this approach for the example under consideration is illustrated in Figure 9.5.

As described in Chapter 6, the TINA-C service architecture models all the features that govern the service life cycle in a TINA-C system in relation to the business model. Using this model, we can say that the access session is used to establish a binding between a service user role and a service provider role. Figure 9.6 illustrates the two access sessions needed for service usage. The first of these is initiated by Mr. Smith to access the VoD service and is directed to the video distributor company. The second access session is initiated between the video distributor company and the video server to enable access to the video service. These access sessions are supported by a communication session and are provided via Internet communications. Telco X is in charge of providing a network connection to support the access session between Mr. Smith and the video distributor company, while telco Y is in charge of supporting the second access session. These access sessions are established between two roles, the *user domain access session* and the *provider domain access session,* with each role contained in its associated domain access session, as shown in Figure 9.6.

The usage sessions shown in Figure 9.7 illustrate cooperation between the retailer (the video server access retailer) and the third-party service provider (the video provider) roles to fulfill the service features requested by the consumer. The usage session takes place between Mr. Smith's user domain usage session and the video server company provider domain usage session.

Further discussion of the communication session used to support the end-to-end service session will be provided within the following section. In our example, the communications session that supports the usage session is established between the application that delivers the video stream and the video consumer premises. The communication session makes use of services from a connectivity provider, whose role is fulfilled by telco Y, to establish and control

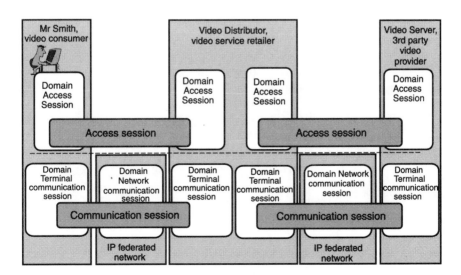

Figure 9.6 Access sessions and related IP communication sessions.

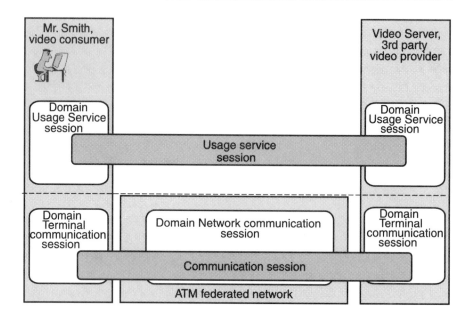

Figure 9.7 Usage session and related ATM communication session.

the network connections between the video server premises and Mr. Smith's premises.

The existence of a session between administrative domains implies that information flows are passing between the respective domains. These information flows arise from the implementation of the computational viewpoint (i.e., a specific computational model), which fulfills the enterprise requirements defined from the enterprise model comprised of the business and session models.

9.2.5   Enterprise Activities

The service logic may be described in terms of enterprise activities and documented via UML collaboration diagrams (as suggested in Chapter 7). We will first introduce a set of naming conventions consistent with Chapter 8 concepts in order to assist in the description of service and management activities.

9.2.5.1   Naming Conventions

To make the example simple, we use a common naming convention within the remainder of this chapter. Names have the following structure: one or more naming components separated by the symbol ".". Naming components may include administrative domains (e.g., X_a representing telco X access network), topological entities (e.g., R_A representing a particular ring), or management functional entities (e.g., fm representing some particular fault management function). Interacting domains are separated by a "-" and enclosed in brackets; for example, [domain1 name - domain2 name]. Manager and agent roles are represented via their domains, which are named by appending the label Mg or Ag, respectively, to some bracketed named entity (e.g., $X_a.Ag$). In a similar fashion, we may represent service roles by their domains with associated labels appended (e.g., Smith.ASR).

9.2.5.2 Service Activity Description

Using the preceding naming conventions, the domain names and definitions associated with the access and usage sessions introduced within the previous section are provided in Table 9.1.

The service is initiated when Mr. Smith decides to access his VoD service; in doing so, he assumes the role of access session requester (Smith.ASR). After the identification and authentication phase, Mr. Smith may select a movie (by clicking on the appropriate movie title) from among those offered by the video distributor. Here the video distributor assumes the role of access session provider (VD.ASP), as illustrated in Figure 9.8. The video distributor, in its role as video access session requester (VD.ASR), opens an access session with the video server (VS.ASP) to gain access to the desired movie on behalf of Mr. Smith. If everything proceeds correctly, an acknowledgment is sent back to Mr. Smith that his request has been received and will be honored.

Table 9.1
Domain Names and Definitions

Domain Name	Domain Definition
Smith.ASR	Mr. Smith domain access session as requester for VoD service access
VD.ASP	VD company domain access session as provider for VoD service access
VD.ASR	VD company domain access session as requester for movie service access
VS.ASP	Video server company domain access session as provider for movie service access
Smith.USC	Mr. Smith domain usage session as consumer for VoD service
VS.USP	Video server company domain usage session as provider for VoD service
VD.CSR	VD company domain communication session as requester for VoD usage
TelcoY.CSP	TelcoY as communication session provider for VoD usage

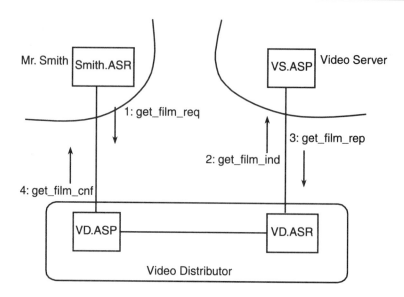

Figure 9.8 Access sessions represented via a UML collaboration diagram.

In honoring the request, as illustrated in Figure 9.9, the video server usage session provider (VS.USP) emits the selected film that is then "consumed" by Mr. Smith (Smith.USC).

In order to support the necessary flow of information between them while the movie is being provided, it is necessary to establish a binding between Mr. Smith's premises and those of the video server. We will now introduce the topological aspects that support this binding before we discuss the detailed communication session that relies upon this information.

9.3 Topology and Network Architecture

Within this section, we illustrate the physical and logical topologies of the networks involved in supporting the VoD service. As noted earlier, service

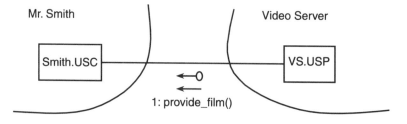

Figure 9.9 Usage sessions represented via a UML collaboration diagram.

realization involves two telecommunications operators, one providing core network and video server and distribution premises access (telco Y) and the other providing residential premises access (telco X). The core network itself is comprised of two rings, each of which is from a different vendor.

9.3.1 Physical Topology

The physical topology of this scenario is illustrated in Figure 9.10. The video consumer, Mr. Smith, has PC, telephone, and television equipment (involving an IP connection, and 64 Kbps and MPEG signals, respectively) at home. As described earlier, Mr. Smith has a contract with an access network operator to supply connectivity for accessing those services and additionally has a contract with the video distributor for the VoD service. He uses his PC and an Internet connection to request video services. The access network operator, telco X, connects the premises equipment with the access network using ADSL technology. The splitter passively directs telephone traffic onto the PSTN, with the remaining video and Internet traffic shown going to the ATM switch via the ADSL modem on two separate ATM trails.

The ATM trails are adapted for carriage on the SDH transport facilities provided in telco Y's SDH core network, which is composed of two interconnected SDH (STM-16) rings, each one provided by a different vendor. Each ring has its own vendor-specific element management system (from the EMS-NE interface perspective) that provides equipment and ring subnetwork management capabilities. The core network is interconnected with telco Y's access network, which provides interconnection with the video server and distributor premises equipment. An STM-4 signal from the SDH core network is sent to an ATM switch at the ingress to telco Y's access network.

Within telco Y's access network, the ATM switch is connected to premises equipment via an SDH signal. As illustrated in Figure 9.10, we see that the ATM switch is being used to enable service access to the desired network (i.e., core network, IP network). Here, an SDH multiplex collects each STM-1 signal coming from the video distributor and video server premises and adapts them into an STM-4 for transport to the ATM switch. We note that the video access units terminate the STM-1 signals carrying the ATM connections and recover the video MPEG signal. There are several video servers that can provide VoD and one video distributor that receives and processes customer requests, relaying them to the appropriate video server.

Obviously, both access networks are very likely to contain more than one switch; however, in the interest of simplicity, we do not make use of more than one switch in this example.

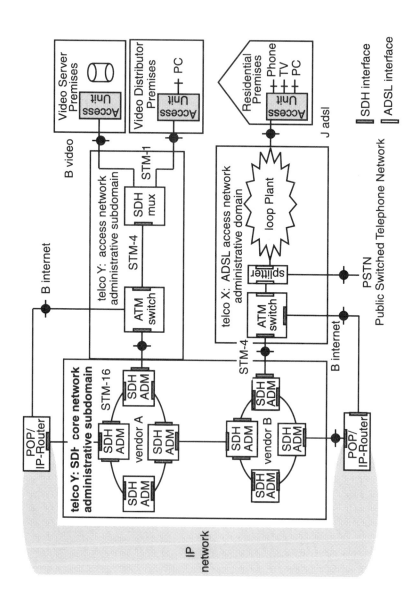

Figure 9.10 Physical topology of VoD service scenario.

9.3.2 Logical Network Architecture

In Section 9.3.1, we have described the physical topology of the network and equipment associated with our example. We will now provide a logical description of the physical network using the functional modeling techniques of Chapter 3, and depicted in Figure 9.11.

For the purposes of this example, we will concentrate on the connections necessary to support the communications sessions that provide the service session of the video distribution business, which is depicted in Figure 9.7. We already decided that this communications session will be delivered using ATM, so the largest layer network dealt with in this scenario is an ATM network (bounded by connection points) supporting the video consumer, video service retailer, and video provider roles. This ATM network is partitioned into subnetworks, which correspond to the access telco and residential and video premises domains of Figure 9.11. Again referring to Figure 9.10, we see that the residential premises subnetwork is connected to the ATM subnetwork of access provider, telco X, via a link supported by ADSL modem technology. The adaptation and termination functions shown are compound functions, which we can expand to show more detail as needed. We note that while the residential premises subnetwork allows for possible flexibility in the home access unit, it is more than likely that this subnetwork will not offer connection flexibility (and

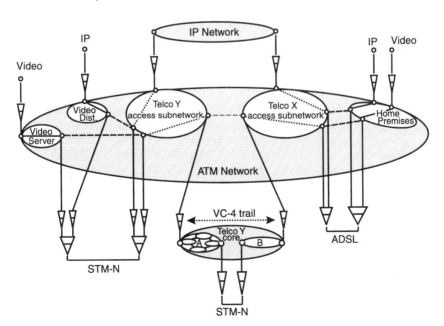

Figure 9.11 Logical network architecture of VoD service scenario.

will thus be degenerate). The video server and video distributor subnetworks are connected to access provider, telco Y, by links supported by VC-4 trails, which are, in turn, carried over a shared SDH resource. We note that the termination functions of these trails are physically located within the ATM switch shown in Figure 9.10. While the ATM part of each access network is modeled by a subnetwork, in this example that subnetwork comprises the single ATM switch.

The access provider domains are interconnected via VC-4 trails supported by the core network provider, telco Y. This VC-4 subnetwork is partitioned into vendor A and vendor B subnetworks, which are interconnected by a link supported over a high-capacity SDH system. Vendor A's subnetwork is depicted as further partitioned into the individual ADMs of Figure 9.10. Since it is not essential to the discussion of this example, we do not depict any further partitioning of vendor B's subnetwork, just as we did not expose the composition of the links within vendor A's subnetwork.

For completeness, the IP network is depicted as a client of the ATM network, allowing IP service to be delivered to ATM network clients. The transport of PSTN service over the loop plant is not depicted since this service plays no role in our example. Again, all adaptation and termination functions shown are expressed in terms of compound functions, which could be expanded to show more detail as needed.

Now that we have a logical model of the network topology, we can turn our attention to a specific connection across the network, which is a cross-section of the topology model. We now consider the composition of a video network connection between the residential consumer and the video producer. This particular communications session supports the service session of the business model and delivers the video program. The video network connection is provided by an ATM trail, composed of the video server subnetwork connection, telco Y access link connection, telco Y access subnetwork connection, telco Y core link connection, telco X access subnetwork connection, telco X access link connection, and finally the home premises subnetwork connection. As shown in Figure 9.12, the telco Y core link connection is provided by a VC-4 trail across the VC-4 layer network.

9.4 Management and Control Aspects

Network management applications are driven by service management requirements that, in this case, are also service requirements, that is, establishment of a stream flow between Mr. Smith's premises and the video server premises. In the same way, service management requirements are affected by network management in the event that a network failure interrupts the service delivery,

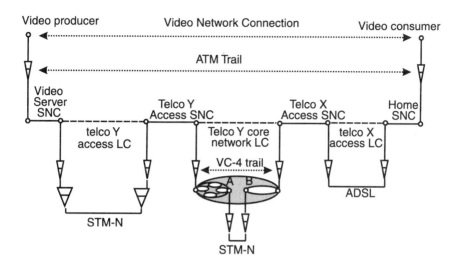

Figure 9.12 Video network connection.

that is, stopping the billing for the VoD service when the failure persists for more than one minute. Within this section, we will focus upon these network-level management aspects, addressing management interactions between telcos X and Y, and intra-telco Y management interactions (TMN X- and Q3 interfaces, respectively, as described within Chapter 2). While we could use either signaling or management protocols to control the communications involved in supporting the VoD service, we only address the use of management protocols in our example. The management systems involved are depicted in Figure 9.13. We note that an engineering solution using signaling protocols could also be developed in a similar manner based upon the same computational model.

9.4.1 Management Activities

Management activities and their associated operations will be expressed using computational models. As previously discussed in Chapter 6, the access and usage sessions described in Section 9.2.4 are supported by communication sessions. Communications sessions are used for two purposes: one for conveying streams between stream producers (the video server) and stream consumers (Mr. Smith), and the other for conveying operations and announcements between operational interfaces (e.g., during the VoD service request phase). In the VoD service architecture, a service session is supported by a communication session, which is in charge of controlling communications over the entire life cycle (i.e., set up, maintenance, release) of the communication session. As

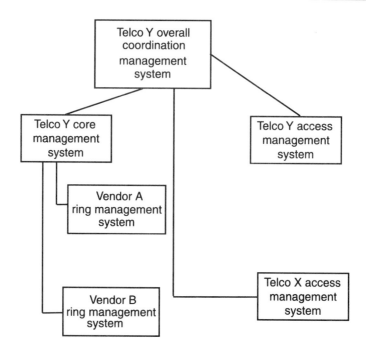

Figure 9.13 Management systems involved in VoD service.

shown in Figure 9.14, these connections may take place within or between premises equipment (terminal connections and subnetwork connections, respectively). Terminal connections are controlled by a terminal communication session manager, while subnetwork connections are controlled by a communication session manager, which additionally has the responsibility of maintaining the end-to-end communication service.

The video network connection shown in Figure 9.12 is controlled by the connection coordinator component. In support of the service session, the connection coordinator computational object will be developed as part of telco Y's access network domain because it is responsible for the overall coordination of communications involved in the VoD service.

9.4.1.1 Decomposition Into Management Domains

The decomposition of the management architecture into management service components is developed from the service requirements, administrative partitioning into enterprise domains (Mr. Smith, video distributor, video server, telco X, and telco Y), and the involved technologies (ADSL, IP, ATM, and SDH). According to the TMN principles described in Chapter 2, these components may be classified according to equipment management, network manage-

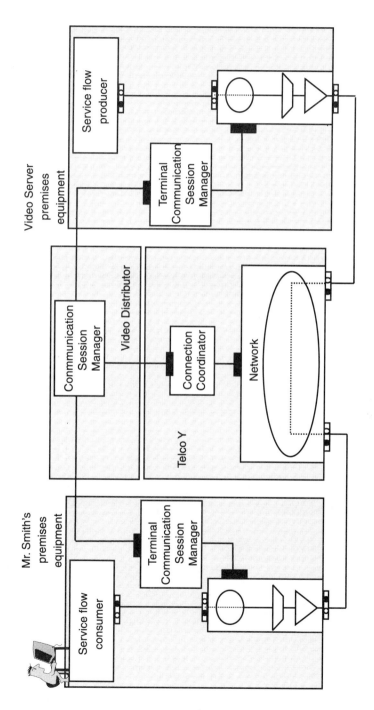

Figure 9.14 Relationship between communication sessions and connections.

ment, and service management. Management service components may be further classified according to the TMN functional areas of fault, configuration, accounting, performance, and security. For example, the obligations "price per view" and "billing schedule" will impact the accounting process, the obligation "identification/authentication" will impact the security management service, and any service failure will affect both the fault and accounting management services. In this example, there is no policy with regard to QoS, so we do not need to consider performance management. Obviously, the configuration management area is mainly involved during the VoD service and network provisioning phases.

We can use the session concept developed in the TINA-C service architecture model to structure the management interactions between the preceding domains. Following TMN terminology, these domains may be qualified as manager or agent domains. Each time an association between management applications is to be established, associated management sessions are created. For example, at the end of each month, the video distributor company will establish a management session with Mr. Smith, telco Y, and the video server to handle accounting management agreements. These interactions are considered to be in the business or service management layer of the TMN layer architecture, as illustrated in Figure 9.15.

The highest level management domain is administrative by nature and is concerned with telco Y's contractual needs (e.g., accounting and security).

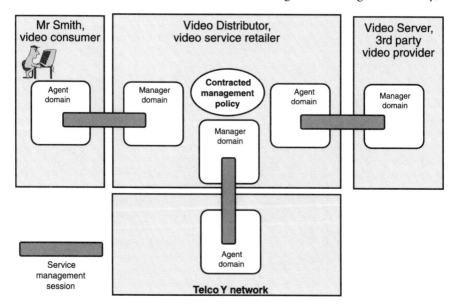

Figure 9.15 Business and service management sessions.

This management domain will delegate other aspects of management to the access management subdomain and the core network management subdomain. The core network management subdomain, in turn, is decomposed into two ring management domains, each of which handles the management of its particular ring subnetwork. Finally, each ring subnetwork management domain is further decomposed into ADM equipment management domains. Telco X and telco Y's access management domains may similarly be partitioned as shown in Figure 9.16.

As we can see, the management domains in this example are organized into a tree structure. While the organization adopted within our example appears quite straightforward, with domains cleanly and hierarchically structured, in the real world management domains may overlap. This may occur, for example, because an operator's internal organizational structure may result in a different domain decomposition for the fault management area from that for the configuration management area. As another example, a management session between telco Y's core network management domain and an SDH ring management domain may mix both network and equipment interactions due to a particular management policy decided upon by the SDH ring equipment vendor. In general, overlapping management domains can result in additional complexity and should be avoided when possible.

The domain names and definitions associated with the management sessions introduced within the previous section are provided in Table 9.2.

9.4.1.2 Engineering Distribution of Management Domains in Management Systems

There are a number of different ways to distribute the management domains introduced in Table 9.2 among the management systems depicted in Figure 9.13. One possibility is to distribute the management domains according to organizational structure. We will consider telco X to have its own management system that includes a number of management domains (i.e., [Y-X].Ag, [X-ADSL].Mg, [X-ATM].Mg) and communicates with telco Y's management system (as illustrated in Figure 9.17 and referring to Table 9.2).

If we consider telco Y's core network management system in more detail, we can imagine the following engineering scenario.

- Telco Y has an overall coordination management system that includes the [Y-X].Mg, TelcoY.CSP, [Y-Y_c].Mg, and [Y-Y_a].Mg management and service domains.

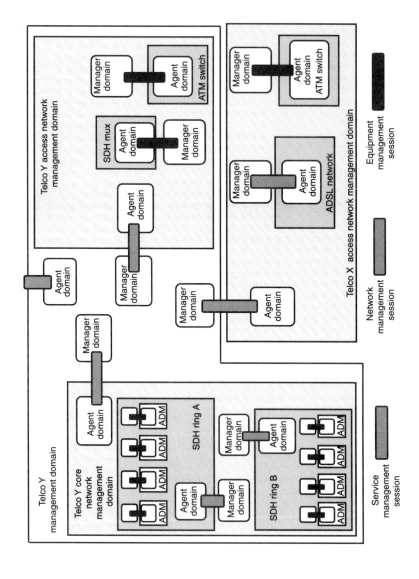

Figure 9.16 Network and equipment management sessions.

Table 9.2
Domain Names and Definitions for Management Sessions

Domain Name	Domain Definition
TelcoY.CSP	Telco Y as communication session provider for VoD usage
[Y-X].Mg	Telco Y as manager in the management session with telco X access network management domain
[Y-X].Ag	Telco X as agent in the management session with telco Y management domain
[X-ADSL].Mg	Telco X access network management domain as manager in the management session with the ADSL management domain
[X-ADSL].Ag	ADSL management domain as agent in the management session with telco X access network management domain
[X-ATM].Mg	Telco X access network management domain as manager in the management session with the ATM switch
[X-ATM].Ag	ATM switch as agent in the management session with telco X access network management domain
[Y-Y_a].Mg	Telco Y as manager in the management session with telco Y access network management domain
[Y-Y_a].Ag	Telco Y access network management domain as agent in the management session with telco Y
[Y_a-SDH].Mg	Telco Y access network management domain as manager in the management session with the SDH multiplex
[Y_a-SDH].Ag	SDH multiplex as agent in the management session with telco Y access network management domain
[Y_a-ATM].Mg	Telco Y access network management domain as manager in the management session with the ATM switch
[Y_a-ATM].Ag	ATM switch as agent in the management session with the telco Y access network management domain
[Y-Y_c].Mg	Telco Y as manager in the management session with telco Y core network management domain
[Y-Y_c].Ag	Telco Y core network management domain as agent in the management session with telco Y
[Y_c-R_A].Mg	Telco Y core network management domain as manager in the management session with SDH Ring A management domain
[Y_c-R_A].Ag	SDH Ring A management domain as agent in the management session with telco Y core network management domain
[R_A-ADM1].Mg	SDH Ring A management domain as manager in the management session with the Add/Drop Multiplex ADM #1
[R_A-ADM1].Ag	Add/Drop Multiplex ADM #1 as agent in the management session with SDH Ring A management domain
[R_A-ADM2].Mg	SDH Ring A management domain as manager in the management session with the Add/Drop Multiplex ADM #2
[R_A-ADM2].Ag	Add/Drop Multiplex ADM #2 as agent in the management session with SDH Ring A management domain
[R_A-ADM3].Mg	SDH Ring A management domain as manager in the management session with the Add/Drop Multiplex ADM #3

Table 9.2 *(continued)*

Domain Name	Domain Definition
[R_A-ADM3].Ag	Add/Drop Multiplex ADM #3 as agent in the management session with SDH Ring A management domain
[R_A-ADM4].Mg	SDH Ring A management domain as manager in the management session with the Add/Drop Multiplex ADM #4
[R_A-ADM4].Ag	Add/Drop Multiplex ADM #4 as agent in the management session with SDH Ring A management domain
[Y_c-R_B].Mg	Telco Y core network management domain as manager in the management session with SDH Ring B management domain
[Y_c-R_B].Ag	SDH Ring B management domain as agent in the management session with telco Y core network management domain
[R_B-ADM1].Mg	SDH Ring B management domain as manager in the management session with the Add/Drop Multiplex ADM #1
[R_B-ADM1].Ag	Add/Drop Multiplex ADM #1 as agent in the management session with SDH Ring B management domain
[R_B-ADM2].Mg	SDH Ring B management domain as manager in the management session with the Add/Drop Multiplex ADM #2
[R_B-ADM2].Ag	Add/Drop Multiplex ADM #2 as agent in the management session with SDH Ring B management domain
[R_B-ADM3].Mg	SDH Ring B management domain as manager in the management session with the Add/Drop Multiplex ADM #3
[R_B-ADM3].Ag	Add/Drop Multiplex ADM #3 as agent in the management session with SDH Ring B management domain
[R_B-ADM4].Mg	SDH Ring B management domain as manager in the management session with the Add/Drop Multiplex ADM #4
[R_B-ADM4].Ag	Add/Drop Multiplex ADM #4 as agent in the management session with SDH Ring B management domain

- Telco Y's overall coordination management system communicates with three distinct management systems: telco X's access network management system, via its inclusion of the [Y-X].Mg management domain; its access network management domain, via its inclusion of the [Y-Y_a].Mg domain; and its core network management domain, via its inclusion of the [Y-Y_c].Mg domain.

- Telco Y's access network management system will then communicate with two subsystems: one for the SDH part of the access network, via its inclusion of the [Y_a-SDH].Mg domain, and one for the ATM part of the access network, via its inclusion of the [Y_a-ATM].Mg domain.

As illustrated in Figure 9.18, telco Y's core network management system interacts with two subsystems, one for each of the SDH rings pertaining to the core network (via its inclusion of the [Y_c-R_A].Mg and [Y_c-R_B].Mg domains).

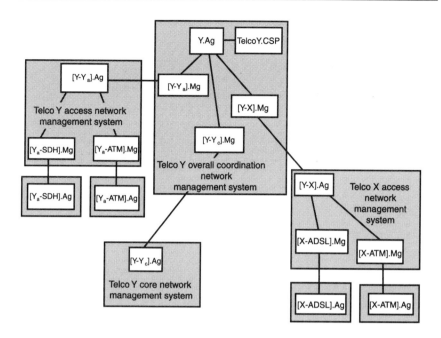

Figure 9.17 Distribution of management domains according to organizational structure.

The management system for each ring interacts with each of its constituent equipments. Thus,

- Ring A's management system includes the management domains [Y_c-R_A].Ag and [R_A-ADM1].Mg, [R_A-ADM2].Mg, [R_A-ADM3].Mg, and [R_A-ADM4].Mg.

- Ring B's management system includes the management domains [Y_c-R_B].Ag and [R_B-ADM1].Mg, [R_B-ADM2].Mg, [R_B-ADM3].Mg, and [R_B-ADM4].Mg.

Finally, each Add/Drop Multiplex includes its own internal management domain (i.e., [R_A-ADM1].Ag, [R_A-ADM2].Ag, [R_A-ADM3].Ag, [R_A-ADM4].Ag or [R_B-ADM1].Ag, [R_B-ADM2].Ag, [R_B-ADM3].Ag, [R_B-ADM4].Ag).

9.4.2 Configuration Management

This section focuses network configuration by providing a sample activity for connection establishment and detailing a particular action that sets up a subnetwork connection across a given subnetwork.

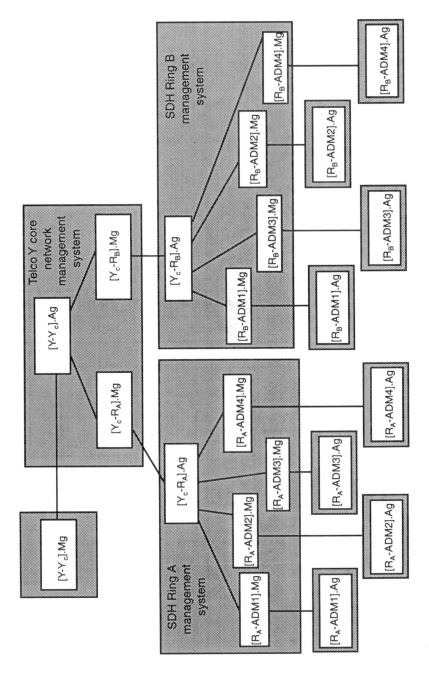

Figure 9.18 Distribution of management domains inside the core network.

9.4.2.1 Computational Activity for Connection Provisioning

Figure 9.19 depicts how interactions occur between the various management domains described in the preceding section in order to provide an end-to-end connection between the video server premises and Mr. Smith's premises. The video distributor, acting as the connectivity service requester (VD.CSR), petitions telco Y to coordinate the establishment of this connection. Telco Y's management system agent (Y.Ag) then communicates with the managers in charge of the various parts of the network connection (i.e., [Y-Y$_a$].Mg for the access part of telco Y's network, [Y-Y$_c$].Mg for the core part of telco Y's network, [Y-X].Mg for the telco X network) and issues requests to their respective configuration management agents (respectively, [Y-Y$_a$].Ag, [Y-Y$_c$].Ag, and [Y-X].Ag).

The process continues until all of the requests are completely cascaded. For example, considering telco Y's core network, requests for connection establishment are sent out to the two ring managers (i.e., [Y$_c$-R$_A$].Mg for Ring A and [Y$_c$-R$_B$].Mg for Ring B). In turn, these ring managers send out requests for connection set-up to the equipment managers, that is, to each add-drop multiplex. The various configuration management sessions involved in provisioning this end-to-end connection are illustrated in Figure 9.19.

9.4.2.2 Connection Set-Up Example

In this section, we present how the connection set-up action described within the previous section may be specfied. In particular, we will describe the associated computational interface specification for connection provisioning in sufficient detail to enable an understanding of how the various parts of the specification are developed. In the interests of clarity, not every specification statement will be reproduced within the main body of the chapter, but more complete specifications will be provided within the appendixes.

The set-up subnetwork connection operation establishes a subnetwork connection between explicitly named termination points representing the A-end and Z-end extremities. It creates a single point-to-point unidirectional or point-to-point bidirectional subnetwork connection object that associates the A-end and Z-end. The client may supply a unique user label for the requested subnetwork connection; if not, the subnetwork provider itself will assign one according to its own policy. If used, a service characteristic specified in the operation input parameters designates a predetermined set of desired server transport parameters. In the event that the subnetwork connection set-up operation's request cannot be satisfied, detailed information will be provided to the service requester regarding the reason(s) the operation failed. The specific reasons why a set-up subnetwork connection operation might fail are enumerated here.

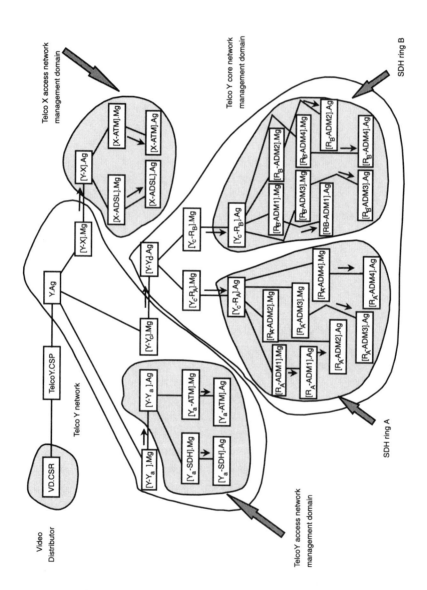

Figure 9.19 Configuration management sessions—UML collaboration diagram.

As discussed in Chapter 4, the precondition of an operation is the system state immediately before the operation is invoked. Exceptions cause the termination of an operation. The preconditions for the set-up subnetwork connection operation indicate the operation will fail if:

- Any of the termination points specified are already involved in a subnetwork connection. This will result in the generation of the exception "subnetworkTerminationPointsConnected."

- Any of the termination points or network termination points are not contained within the domain of the subnetwork. This will result in the generation of the exception "incorrectSubnetworkTermination-Points."

Again, as discussed in Chapter 4, the postcondition of an operation reflects the state of the system immediately after the operation completes. The postconditions for the set-up subnetwork connection operation indicate the operation will fail if:

- Any of the subnetwork connection input parameters could not be validated by the server.

- The value of the userLabel of the SubnetworkConnection is not unique within the domain of the containing subnetwork. This will result in the generation of the exception "userLabelInUse."

We will now introduce an explicit example of a set-up subnetwork connection operation. The topology considered in this example is illustrated in Figure 9.20, where we identified some particular termination points among which connections may be established. We note that each of the management systems described earlier in this chapter will be able to use the same generic set-up subnetwork connection operation in establishing the end-to-end connection. This will become obvious in the following example.

Let us consider establishing a connection from the video distributor to Mr. Smith's premises. For simplicity, we will only consider that portion of the connection between connection points Y_a.ATM.P1 and X_a.ATM.P3. In order to establish this ATM connection, we must establish a contiguous ATM link connection between Y_a.ATM.P1 and X_a.ATM.P2 and an ATM subnetwork connection between X_a.ATM.P2 and X_a.ATM.P3. Establishing this subnetwork connection is accomplished by telco X via the set-up subnetwork connection operation with the appropriate parameters related to telco X's ATM subnetwork. Establishing the link connection within telco Y's administrative domain might

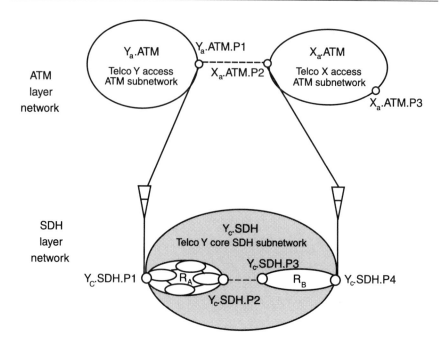

Figure 9.20 Connection set-up scenario.

require establishing a trail in the SDH server layer network in the case where no spare capacity is available (through the use of link connection, adaptation, and trail management services interacting together). In such a case, the trail management service will be in charge of setting up a trail. This will require establishing a subnetwork connection between $Y_c.SDH.P1$ and $Y_c.SDH.P4$ inside telco Y's core network.

Referring to Figure 9.20 and considering only the core network, the subnetwork connection set-up operation request is to set up an SDH subnetwork connection between the $Y_c.SDH.P1$ (the A-end) and $Y_c.SDH.P4$ (the Z-end) connection points.

The telco Y connection performer receiving the request would then compute a route between these two connection points and send the following requests, where $Y_c.SDH.R_A$, $Y_c.SDH.R_B$, $Y_c.SDH.R_A.P1$, $Y_c.SDH.R_A.P2$, $Y_c.SDH.R_B.P3$, $Y_c.SDH.R_B.P4$, *bidirectional,* "*Smith_connection*", and *rsc* (requested service characteristics) are the actual values passed as parameters in the operation call:

- ssccSetupSubnetworkConnection ($Y_c.SDH.R_A$, $Y_c.SDH.R_A.P1$, $Y_c.$-SDH.$R_A.P2$, bidirectional, "Smith_connection", rsc);

- ssccSetupSubnetworkConnection (Y$_c$.SDH.R$_B$, Y$_c$.SDH.R$_B$.P3, Y$_c$.SDH.R$_B$.P4, bidirectional, "Smith_connection", rsc).

Referring to Figure 9.21, each ring connection performer then (for example) sends the following request to the ADM connection performers:

- ssccSetupSubnetworkConnection (Y$_c$.SDH.R$_A$.ADM1, Y$_c$.SDH.-P1, Y$_c$.SDH.R$_A$.P1, bidirectional, "Smith_connection", rsc);
- ssccSetupSubnetworkConnection (Y$_c$.SDH.R$_A$.ADM2, Y$_c$.SDH.-R$_A$.P2, Y$_c$.SDH.R$_A$.P3, bidirectional, "Smith_connection", rsc);
- ssccSetupSubnetworkConnection (Y$_c$.SDH.R$_A$.ADM3, Y$_c$.SDH.-R$_A$.P4, Y$_c$.SDH.P2, bidirectional, "Smith_connection", rsc).

Now that we explained how the set-up subnetwork connection operation is used in an application, we provide an annotated and abbreviated version of the associated semiformal specification (comments are designated by "--" followed by italic text). This specification is based upon the computational viewpoint structure introduced within Appendix 4C. The interested reader can find the complete semiformal specification in Appendix 9A.

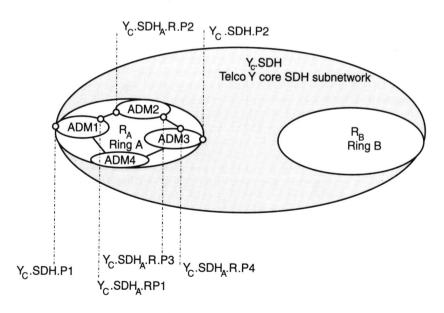

Figure 9.21 Ring set-up connection scenario.

COMPUTATIONAL_INTERFACE simpleSncPerformerIfce {
-- *the computational interface contains the set-up and release operations*
    OPERATION <ssccSetupSubnetworkConnection>;
                <ssccReleaseSubnetworkConnection>;
}
-- *we only illustrate the set-up operation*
OPERATION ssccSetupSubnetworkConnection {
    INPUT_PARAMETERS
    -- *The subnetwork is an interface reference to the subnetwork being*
    -- *operated on. This allows remote access to any subnetwork in*
    -- *order to enable the potential for distribution of the interfaces.*
    -- *ssccSnIfce is a type of query interface (the minimal one for each*
    -- *information object) and it allows access to information related to*
    -- *the subnetwork object.*

        subnetwork : SubnetworkId ::= (ssccSnIfce);
        snpa : SnTPId ::= (snTPIfce);
        -- *The A- and Z-end passed as interface references will be matched to an snTP*
        -- *interface that can provide access to information from either a Trail Termination*
        -- *Point or a Connection Termination Point.*
        snpz : SnTPId ::= (snTPIfce);
        dir : Directionality;
        -- *desired directionality passed as a value*
        suppliedUserLabel : UserLabel ;
        serviceCharacteristics: CharacteristicsId ::= (serviceCharacteristicsIfce) ;

    OUTPUT_PARAMETERS
        newSNC : SNCId ::= (sncIfce) ;
        -- *interface reference of the created subnetwork connection*
        agreedUserLabel : UserLabel ;
        -- *and its label.*

    RAISED_EXCEPTIONS
    -- *These exceptions match the failures of the pre- and postconditions.*
    -- *Here the data types of the exceptions are defined. We will only*
    -- *elaborate the subnetworkTerminationPointsConnected exception.*

        subnetworkTerminationPointsConnected : SEQUENCE OF SnTPId;

    BEHAVIOR

        PARAMETER_MATCHING
        -- *Gives the binding between the declared parameter and the*
        -- *corresponding part (object or attribute) of the information*
        -- *viewpoint, which gives semantics to this parameter (it can be*
        -- *thought of as a semantic type instead of a syntaxic type).*
            subnetwork: < INFORMATION OBJECT: ssccSubnetwork>;
            -- *the subnetwork parameter is bound to the ssccSubnetwork*

-- *information object and then will have all its characteristics.*
snpa : < INFORMATION OBJECT: ssccSubnetworkTP>;
snpz : < INFORMATION OBJECT: ssccSubnetworkTP>;
dir : <INFORMATION ATTRIBUTE: directionality >;
newSNC : < INFORMATION OBJECT: ssccSubnetworkConnection>;
suppliedUserLabel : < INFORMATION ATTRIBUTE: userLabel >;
agreedUserLabel : < INFORMATION ATTRIBUTE: userLabel >;
serviceCharacteristics : <INFORMATION OBJECT
       involvedServiceCharacteristics >;

-- *We note that in the following conditions "bold type" refers to*
-- *parameters while "italic type" refers to relationships and information*
-- *objects or attributes.*

PRE_CONDITIONS
-- *Preconditions are defined as a list of predicates. The definition of the*
-- *predicates constrains the parameters through the use of information*
-- *relationships that are verified by their bound information objects.*

inv_belongingpoints
"The **snpA** and **snpZ** must refer to *element* in the *subnetworkIsDelimitedBy*
relationship where **subnetwork** refers to *ssccsubnetwork.*
-- *The parameters snpA and snpB are bound to a ssccSubnetworkTP*
-- *information object type, and the parameter subnetwork is bound to a*
-- *ssccSubnetwork information object. These three information objects*
-- *shall be related through a subnetworkIsDelimitedBy information*
-- *relationship.*

inv_pointANotConnected
"The **snpA** shall not refer to any *A-end* of a
subnetworkConnectionIsTerminatedByPointToPoint **relationship**."

POST_CONDITIONS
inv_connectedPoints
"The **newSNC**, **snpa** and **snpz** must refer respectively to *transportEntity, A_end* and
Z_end in a *subnetworkConnectionIsTerminatedByPointToPoint* relationship."
-- *The parameter newSNC is bound to the SNC information object, which plays*
-- *a transportEntity role in the subnetworkConnectionIsTerminatedByPointToPoint*
-- *relationship where the A_and Z_end are played by the corresponding snpa*
-- *and snpz parameters.*

EXCEPTIONS
-- *In this section explicit behavior is assigned for each pre- and post-*
-- *condition that is not satisfied. The data type of each exception*
-- *was already defined in the RAISED_EXCEPTIONS section.*
-- *We only present a single condition as an example. The invariant is*
-- *already defined in the preconditions clause, and the exception is defined*
-- *in the exceptions clause. The pre- or postcondition will give the reason*

-- why the exception was raised.

    IF PRE_CONDITION <inv_pointANotConnected> NOT_VERIFIED
      RAISE_EXCEPTION subnetworkTerminationPointsConnected ;
  ;
}

9.4.2.3 UML Diagrams for Connection Setup

The information object class diagram illustrated in Figure 9.22 presents the informational subsystem that is being manipulated and affected by the ssccSetupSubnetworkConnection operation. It represents G.853-01 object classes and their specialization in G.853-02. As can be observed, attributes related to the set up of a simple subnetwork connection (sscc) have been defined in the sscc-related object classes.

    The class diagram in Figure 9.23 represents the relationship between the information specification and the precondition related to the operation ssccSetupSubnetworkConnection. It is based upon the information entities introduced in Figure 9.22 but also captures parameter matching and the predicates forming the precondition. Parameters of the operation ssccSetupSubnetworkConnection are depicted as UML notes (i.e., a rectangle having its northeast corner folded), and their matching to information object classes or attributes is represented via dashed lines between the two. Each predicate forming the precondition is captured in a note and related to the information objects, relationships, and parameters that are implied in it.

    The class diagram in Figure 9.24 represents the relationship between the information specification and the postcondition related to the operation ssccSetupSubnetworkConnection. The diagrammatic conventions are the same as described previously. New relationships appear because the informational subsystem has changed between the "before" state (precondition) and the "after" state (postcondition).

9.4.3 Fault Management

We now demonstrate how an equipment fault can lead to the generation of management data flows. This is shown in Figure 9.25, which illustrates an example where an existing connection fails inside ADM4 of Ring A. When this occurs, an equipment-level notification is sent from ADM4 to Ring A's manager, which transforms it into "subnetwork"-level information and issues a message "subnetwork connection broken" to the telco Y core network manager ([Y-Y$_c$].Ag). In turn, telco Y's core network manager forwards the notification (without change) to telco Y's central fault management agent system (Y.Ag). Knowing that this failed connection was intended to support a service, it sends

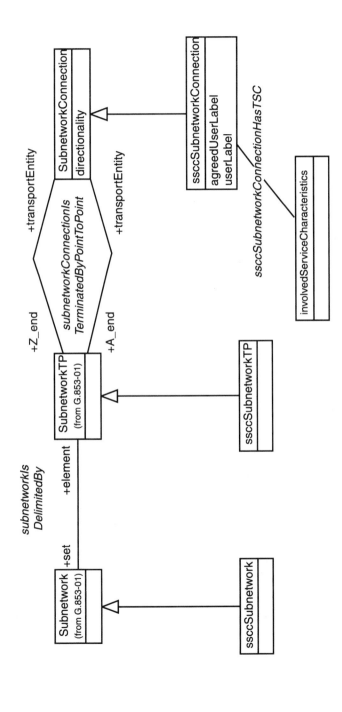

Figure 9.22 Information object class for connection set-up scenario.

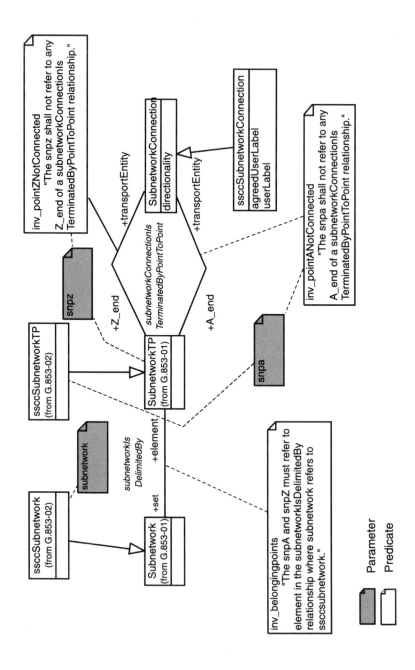

Figure 9.23 Representation of the ssccSetupSubnetworkConnection precondition.

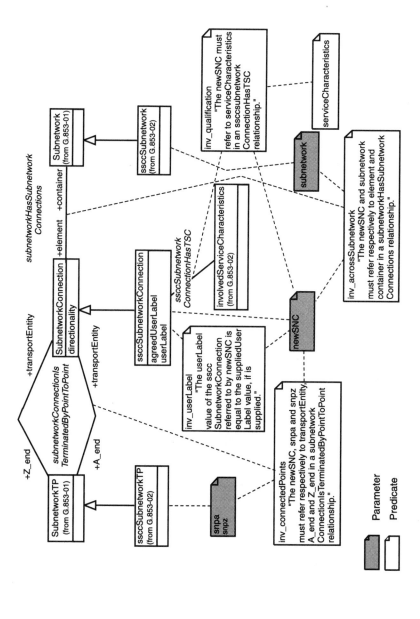

Figure 9.24 Representation of the ssccSetupSubnetworkConnection postcondition.

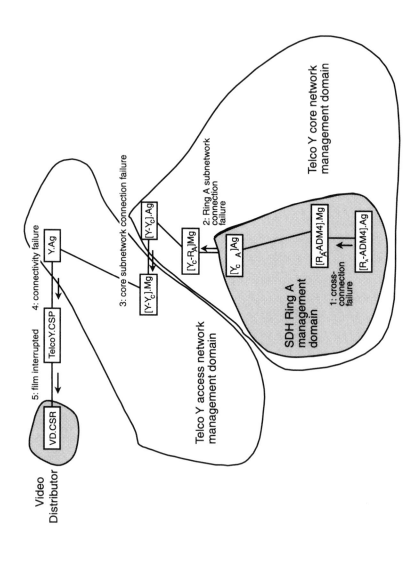

Figure 9.25 Fault management sessions represented via a UML collaboration diagram.

out a notification "connectivity failure" to the telco Y connectivity service provider, which in turn informs the video distributor connectivity service requester (VD.CSR) that the movie has been interrupted.

9.4.4 Engineering Interfaces

In this section, we illustrate how the mapping of the computational specification of subnetwork connection setup can be mapped to a GDMO engineering interface specification and how to translate this to a CORBA IDL engineering interface specification. We note that the following subsections assume more detailed knowledge of GDMO and CORBA IDL on the part of the reader.

9.4.4.1 Application to GDMO

The following GDMO specification is a possible CMIP interface engineering specification corresponding to the realization of the simpleSncPerformerIfce computational interface described in the preceding section. We assign to the computational interface a GDMO managed object class (simpleSubnetwork-ConnectionPerformer), which is defined by a single mandatory package containing an action. This action, setupSnc, is the result of the mapping of the computational operation ssccSetupSubnetworkConnection onto GDMO. Recalling the GDMO template contained in Appendix 2B, we now provide an annotated and abbreviated GDMO specification of the simpleSubnetwork-ConnectionPerformer managed object class (comments are designated by "--" followed by italic text). The interested reader may find the complete specification in Appendix 9B.

```
simpleSubnetworkConnectionPerformer MANAGED OBJECT CLASS
    DERIVED FROM "Recommendation X.721":top; -- the inheritance parent object
        CHARACTERIZED BY
        simpleSubnetworkConnectionPerformerPackage PACKAGE
            BEHAVIOR
                simpleSubnetworkConnectionPerformerBehavior BEHAVIOR
                DEFINED AS "
                    This managed object class is the engineering specification of the
                    computational interface simpleSncPerformerIfce.
                ";;
            ACTIONS -- the only action on this object
                setupSnc; ;
    REGISTERED AS {g85501MObjectClass ..}; -- every GDMO template is registered
```

The definition of the following action is an engineering specification of the computational operation ssccSetupSubnetworkConnection. The input parameters (INPUT_PARAMETERS) of the computational operation are mapped on to the ASN.1 type SetupSncInformation in GDMO in the "WITH INFORMATION SYNTAX" construct. Similarly, the output parameters (OUTPUT_PARAMETERS) of the computational operation are mapped on to the ASN.1 type SetupSncResult in GDMO in the "WITH REPLY SYNTAX construct." The exceptions (RAISED_EXCEPTIONS) that are part of the signature of the computational operation are mapped onto the GDMO parameters in the "PARAMETERS" construct. A one-to-one correspondence exists between computational exceptions and GDMO engineering parameters.

```
setupSnc ACTION
    BEHAVIOR
        setupSncBehavior BEHAVIOR
            DEFINED AS "
                This action is used to set up a Subnetwork Connection between subnetwork
                termination points.

                The behavior of this action is specified in Simple Subnetwork Connection
                Configuration, clause A.2.1 <G.854-01:OPERATION, setupSnc >
                ";;
    MODE CONFIRMED; -- this specifies protocol characteristics
    PARAMETERS
    -- these parameters are examples of computational exceptions
        subnetworkTerminationPointsConnected,
    WITH INFORMATION SYNTAX
    -- this clause indicates the ASN.1 syntax for the operation parameters
        G85501-ASN1TypeModule.SetupSncInformation;
    WITH REPLY SYNTAX
    -- while this gives response syntax
        G85501-ASN1TypeModule.SetupSncResult;
REGISTERED AS {g85501Action ...};
```

The following example describes an illustrative parameter and associates an ASN.1 type definition to it:

```
subnetworkTerminationPointsConnected PARAMETER
        CONTEXT ACTION-REPLY;
        WITH SYNTAX G85501-ASN1TypeModule.SNTPList;
REGISTERED AS {g85501SpecificError ..};
```

We now provide the ASN.1 type definition module below, which is essentially the input (SetupSncInformation) and output parameters (SetupSncResult) type definition. References to already defined types are captured in the "IMPORTS" construct.

ASN.1 Module
G85501-ASN1TypeModule {itu-t recommendation g(7) gntm(85501) informationModel(0)
asn1Modules(2)
asn1TypeModule(0)}
DEFINITIONS IMPLICIT TAGS ::=
BEGIN
-- *Exports everything; full GDMO not included here*
IMPORTS

    Directionality, UserLabel
       FROM ASN1DefinedTypesModule
       {ccitt recommendation m(13) gnm(3100) informationModel(0) asn1Modules(2)
          asn1DefinedTypesModule(0)}

    ObjectInstance
       FROM CMIP-1 {joint-iso-ccitt ms(9) cmip(1) modules(0) protocol(3)}

CharacteristicsId : := ObjectInstance

SetupSncInformation ::= SEQUENCE {
        subnetwork   subnetworkId,
        snpa   SnTPId,
        snpz   SnTPId,
        dir   Directionality,
        suppliedUserLabel   [0] UserLabel OPTIONAL,
        serviceCharacteristics CharacteristicsId
        }

SetupSncResult ::= SEQUENCE {
        newSNC   SNCId,
        agreedUserLabel   UserLabel
        }

SNCId : := ObjectInstance

SnTPId : := ObjectInstance

SNTPList : := SEQUENCE OF SnTPId

SubnetworkId : := ObjectInstance

END

9.4.4.2 Application to IDL

In this section, we provide a possible IDL engineering specification corresponding to the realization of the simpleSncPerformerIfce computational interface

as described earlier. This IDL is derived from the GDMO specification defined in the preceding section by applying the translation rules introduced within Chapter 8. We note that a direct translation of the computational operation signature specified in Section 9.4.2.2 to IDL, which we could have provided, would probably have resulted in a different IDL specification. For the translation approach we are using, the SimpleSubnetworkConnectionPerformer IDL interface is assigned to the simpleSubnetworkConnectionPerformer computational interface. The action, setupSnc, is the result of the mapping of the computational operation ssccSetupSubnetworkConnection onto the IDL. For translations from GDMO to IDL, the comments generated as part of the translation from GDMO to IDL are denoted by the "//" comment prefix. In this text, we use the prefix "//**" with italic text to denote explanatory annotation. While this example employs manual translation, tools are available that allow automatic translation. Analogous to the other specification examples, this IDL specification is abbreviated; the interested reader can find the complete specification in Appendix 9C.

```
//** Existing modules are imported using the "include" construct. It is customary to follow
//** this with comments indicating the modules imported from this file
#include "X711CMI.idl"
// ASN.1 Module name: CMIP-1
// ASN.1 Module OID: 2.9.1.0.3
// ASN.1 Module nickname: X711CMI
//** in what follows, the imported modules have been removed to shorten this section
#include "X721.idl"
#include "M.3100.idl"
#include <JIDM.idl>
#include <OSIMgmt.idl>
#pragma prefix "jidm.org"


module G85501ASN {
    //** this defines several object instance types
    typedef X711CMI::ObjectInstanceType CharacteristicsIdType, SNCIdType,
    SnTPIdType, SubnetworkIdType;

    //** this defines the sequence of sub net TP id's, SnTPListType
    typedef sequence<SnTPIdType> SnTPListType;
    //** these type definitions provide the action input and result structures
    struct SetupSncInformationType
    {
      SubnetworkIdType subnetwork;
      //** format here is "type" "name" pairs
      SnTPIdType snpa, snpz;
      //** so this defines snpa and snpz of type SnTPIdType
      M3100ASN::DirectionalityType dir;
```

```
//** a union expresses variable data that depends on the switch element and its values
union UserLabelTypeOpt switch (boolean)
{case TRUE: M3100ASN::UserLabelType dir;};
CharacteristicsIdType serviceCharacteristics
};

struct SetupSncResultType
{
SNCIdType newSNC;
M3100ASN::UserLabelType agreedUserLabel
};
}
//** this module provides type definitions for the exceptions described in the GDMO
//** PARAMETERS clause
module G85501SSC {
typedef G85501ASN::SNTPListType subnetworkTerminationPointsConnectedType;
//** not all type definitions are shown here
//** note the import of the CMIS protocol Object Instance
// simpleSubnetworkConnectionPerformer
//** this is the interface created from the computational interface specification. Comments trace
//** from which parts of the GDMO specification these interface specifications are derived
//** from
interface SimpleSubnetworkConnectionPerformer:X721::Top {
// simpleSubnetworkConnectionPerformerPackage
// This managed object class is the engineering specification of the computational
// interface simpleSncPerformerIfce
// ACTION: setupSnc
//** at last, the setupSnc action! The format is again "type" "name" and the type
//** is imported from the G85501ASN module defined earlier
G85501ASN::SetupSncResultType setupSnc (
in G85501ASN::SetupSncInformationType actionInfo) //** input parameters
raises (ACTION_ERRORS, UsingMR) //** exception handling
//** In the GDMO specification, the ACTION template has a WITH REPLY SYNTAX
//** clause. Therefore, the action may result in multiple replies. Then, according to the
//** translation rules described in Chapter 8, the mapped operation has a return
//** type that is the translation of the WITH REPLY SYNTAX clause and may raise an
//** user exception named UsingMR to indicate if it is using multiple replies.
//** The ACTION_ERRORS macro is defined in and imported from the file
//** "OSIMgmt.idl." This macro defines IDL exceptions for the whole CMIP and
//** ROSE basic set of errors, allowing any invoker of an IDL operation to handle
//** CMIP errors through exceptions; e.g., InvalidScope, Processing Failure.
}
```

9.5 Summary

In this chapter, we built upon the concepts introduced throughout the book to demonstrate how to create a cohesive service specification that may be translated into associated network and equipment management specifications. As explained in Chapter 6, we employed a unified modeling approach, starting from service requirements and ending with the interface specifications for managing equipment. We used the techniques of Chapter 3 to capture the network architecture; those of Chapter 4 to capture the management aspects; the unified approach of Chapter 6 to establish the relationship between services and transport aspects, including their control and management; and those of Chapter 7 to express these in the UML notation. We showed how to derive a specific engineering realization from a technology-independent specification and illustrated a GDMO/ASN.1 to IDL translation as discussed in Chapter 8 in the context of interdomain management. Thus, this chapter illustrated how the usage of model-based approaches can enable the development of a set of integrated requirements in a distributed world of heterogeneous technologies, network and equipment architectures, management domains, and operator policies.

References

[1] ITU-T Y100, "General Review of Global Information Infrastructure," June 1998.

[2] ITU-T Y110, "Global Information Infrastructure Principles and Framework," June 1998.

[3] ITU-T Y120, "Global Information Infrastructure Scenarios and Methodology," June 1998.

Appendix 9A
Semiformal Specification for simpleSncPerformerIfce

```
COMPUTATIONAL_INTERFACE simpleSncPerformerIfce {
    OPERATION <ssccSetupSubnetworkConnection>;
             <ssccReleaseSubnetworkConnection>;
}

OPERATION ssccSetupSubnetworkConnection {

    INPUT_PARAMETERS
        subnetwork : SubnetworkId ::= (ssccSnIfce);
        snpa : SnTPId ::= (snTPIfce);
        snpz : SnTPId ::= (snTPIfce);
        dir : Directionality;
        suppliedUserLabel : UserLabel ;
        serviceCharacteristics: CharacteristicsId ::= (serviceCharacteristicsIfce) ;

    OUTPUT_PARAMETERS
```

newSNC : SNCId ::= (sncIfce) ;
agreedUserLabel : UserLabel ;

RAISED_EXCEPTIONS
    invalidTransportServiceCharacteristics: NULL;
    incorrectSubnetworkTerminationPoints : SEQUENCE OF SnTPId;
    subnetworkTerminationPointsConnected : SEQUENCE OF SnTPId;
    failure : Failed;
    wrongDirectionality : Directionality;
    userLabelInUse : UserLabel;

BEHAVIOR

SEMI_FORMAL
    PARAMETER_MATCHING
        subnetwork: < INFORMATION OBJECT: ssccSubnetwork>;
        snpa : < INFORMATION OBJECT: ssccSubnetworkTP>;
        snpz : < INFORMATION OBJECT: ssccSubnetworkTP>;
        dir : <INFORMATION ATTRIBUTE: directionality >;
        newSNC : < INFORMATION OBJECT: ssccSubnetworkConnection>;
        suppliedUserLabel : < INFORMATION ATTRIBUTE: userLabel >;
        agreedUserLabel : < INFORMATION ATTRIBUTE: userLabel >;
        serviceCharacteristics : < INFORMATION OBJECT
involvedServiceCharacteristics >;

PRE_CONDITIONS

inv_belongingpoints
"The **snpA** and **snpZ** must refer to *element* in the *subnetworkIsDelimitedBy*
relationship where **subnetwork** refers to *ssccsubnetwork.*
inv_pointANotConnected
"The **snpA** shall not refer to any *A-end* of a
subnetworkConnectionIsTerminatedByPointToPoint **relationship.**"
inv_pointZNotConnected
"The **snpZ** shall not refer to any *Z-end* of a
subnetworkConnectionIsTerminatedByPointToPoint **relationship.**"

POST_CONDITIONS

inv_connectedPoints
"The **newSNC, snpa** and **snpz** must refer respectively to *transportEntity, A_end* and
Z_end in a *subnetworkConnectionIsTerminatedByPointToPoint* relationship."
inv_qualification
"The **newSNC** must refer to *transportQualified* in an
ssccsubnetworkConnectionHasTSC relationship."
inv_acrossSubnetwork
"The newSNC and subnetwork must refer respectively to *element* and *container* in a
subnetworkHasSubnetworkConnections relationship."

inv_userLabel
"The *userLabel* value of the *ssccSubnetworkConnection* referred to by **newSNC** is equal to the **suppliedUserLabel** value, if it is supplied."

EXCEPTIONS

IF PRE_CONDITION <inv_ belongingpoints > NOT_VERIFIED
    RAISE_EXCEPTION incorrectSubnetworkTerminationPoints ;
IF PRE_CONDITION <inv_ pointANotConnected > NOT_VERIFIED
    RAISE_EXCEPTION subnetworkTerminationPointsConnected ;
IF PRE_CONDITION <inv_ pointZNotConnected > NOT_VERIFIED
    RAISE_EXCEPTION subnetworkTerminationPointsConnected ;
IF POST_CONDITION <inv_ connectedPoints > NOT_VERIFIED
RAISE_EXCEPTION failure ;
IF POST_CONDITION <inv_ qualification > NOT_VERIFIED
RAISE_EXCEPTION failure ;
IF POST_CONDITION <inv_ acrossSubnetwork > NOT_VERIFIED
RAISE_EXCEPTION failure ;
IF POST_CONDITION <inv_ userLabel> NOT_VERIFIED RAISE_EXCEPTION
userLabelInUse ;
    ;
}

Appendix 9B
GDMO Specification:
SimpleSubnetworkConnectionPerformer Managed Object Class

simpleSubnetworkConnectionPerformer MANAGED OBJECT CLASS
    DERIVED FROM «Recommendation X.721»:top;
        CHARACTERIZED BY
        simpleSubnetworkConnectionPerformerPackage PACKAGE
            BEHAVIOR
                simpleSubnetworkConnectionPerformerBehavior BEHAVIOR
                DEFINED AS "
                    This managed object class is the engineering specification of the
                    computational interface simpleSncPerformerIfce.
                ";;
            ACTIONS
                setupSnc; ;
REGISTERED AS {g85501MObjectClass ..};

setupSnc ACTION
    BEHAVIOR
        setupSncBehavior BEHAVIOR
            DEFINED AS "

                This action is used to set up a Subnetwork Connection between subnetwork termination points.

                The behavior of this action is specified in *Simple Subnetwork Connection Configuration,* clause A.2.1 <G.854-01:OPERATION, setupSnc >
                ";;
    MODE CONFIRMED;
    PARAMETERS
        invalidTransportServiceCharacteristics,
        incorrectSubnetworkTerminationPoints,
        subnetworkTerminationPointsConnected,
        failure,
        wrongDirectionality,
        userLabelInUse ;
    WITH INFORMATION SYNTAX
        G85501-ASN1TypeModule.SetupSncInformation;
    WITH REPLY SYNTAX
        G85501-ASN1TypeModule.SetupSncResult;
REGISTERED AS {g85501Action ...};

failure PARAMETER
    CONTEXT   ACTION-REPLY;
    WITH SYNTAX   G85501-ASN1TypeModule.Failed;
REGISTERED AS {g85501SpecificError ..};

incorrectSubnetworkTerminationPoints PARAMETER
    CONTEXT   ACTION-REPLY;
    WITH SYNTAX   G85501-ASN1TypeModule.SNTPList;
REGISTERED AS {g85501SpecificError ..};

invalidTransportServiceCharacteristics PARAMETER
    CONTEXT   ACTION-REPLY;
    WITH SYNTAX   G85501-ASN1TypeModule.NULL;
REGISTERED AS {g85501SpecificError ..};

subnetworkTerminationPointsConnected PARAMETER
    CONTEXT   ACTION-REPLY;
    WITH SYNTAX   G85501-ASN1TypeModule.SNTPList;
REGISTERED AS {g85501SpecificError ..};

userLabelInUse PARAMETER
    CONTEXT   ACTION-REPLY;
    WITH SYNTAX   UserLabel;
REGISTERED AS {g85501SpecificError ..};

wrongDirectionality PARAMETER
        CONTEXT   ACTION-REPLY;
        WITH SYNTAX   Directionality;
REGISTERED AS {g85501SpecificError ..};

ASN.1 Module
G85501-ASN1TypeModule {itu-t recommendation g(7) gntm(85501) informationModel(0)
asn1Modules(2)
asn1TypeModule(0)}
DEFINITIONS IMPLICIT TAGS ::=
BEGIN
-- *EXPORTS everything*
IMPORTS

    Directionality, Failed, UserLabel
        FROM ASN1DefinedTypesModule
        {ccitt recommendation m(13) gnm(3100) informationModel(0) asn1Modules(2)
                asn1DefinedTypesModule(0)}
    ObjectInstance
        FROM CMIP-1 {joint-iso-ccitt ms(9) cmip(1) modules(0) protocol(3)}

    SetupSncInformation ::= SEQUENCE {
        subnetwork   subnetworkId,
        snpa   SnTPId,
        snpz   SnTPId,
        dir   Directionality,
        suppliedUserLabel   [0] UserLabel OPTIONAL,
        serviceCharacteristics CharacteristicsId
        }

CharacteristicsId : := ObjectInstance

SetupSncResult ::= SEQUENCE {
        newSNC   SNCId,
        agreedUserLabel   UserLabel
        }

CharacteristicsId : := ObjectInstance

SNCId : := ObjectInstance

SnTPId : := ObjectInstance

SNTPList : := SEQUENCE OF SnTPId

SubnetworkId : := ObjectInstance
END

Appendix 9C
IDL Specification: SimpleSubnetworkConnectionPerformer
IDL Interface

```
#include "X711CMI.idl"
// ASN.1 Module name: CMIP-1
// ASN.1 Module OID: 2.9.1.0.3
// ASN.1 Module nickname: X711CMI


#include "X721.idl"
// ASN.1 Module name: X.721
// ASN.1 Module OID: 2.9.3.2.2.0
// ASN.1 Module nickname: X721


#include "M.3100.idl"
// ASN.1 Module name: M3100- ASN1DefinedTypes
// ASN.1 Module OID: 2.13.3100.0.2.0
// ASN.1 Module nickname: M3100ASN


#include <JIDM.idl>
#include <OSIMgmt.idl>
#pragma prefix "jidm.org"


// ASN.1 Module name: G85501-ASN1TypeModule
// ASN.1 Module OID: 2.7.85501.0.2.0
// ASN.1 Module nickname: G85501ASN
module G85501ASN {
        typedef X711CMI::ObjectInstanceType CharacteristicsIdType, SNCIdType,
        SnTPIdType, SubnetworkIdType;

        typedef sequence<SnTPIdType> SnTPListType

        struct SetupSncInformationType
        {
          SubnetworkIdType subnetwork;
          SnTPIdType snpa, snpz;
          M3100ASN::DirectionalityType dir;
          union UserLabelTypeOpt switch (boolean)
             {case TRUE: M3100ASN::UserLabelType dir;};
          CharacteristicsIdType serviceCharacteristics
        };

        struct SetupSncResultType
        {
          SNCIdType newSNC;
          M3100ASN::UserLabelType agreedUserLabel
```

```
    };
}
//ASN.1 Module name: G85501-SSC
//ASN.1 Module OID: 2.7.85501.0.2.1
// ASN.1 Module nickname: G85501SSC
module G85501SSC {


    typedef G85501ASN::FailedType failureType;
    // ACTION-REPLY


    typedef G85501ASN::SNTPListType
            incorrectSubnetworkTerminationPointsType;
    //ACTION-REPLY


    typedef G85501ASN::NULLType
            invalidTransportServiceCharacteristicsType;
    //ACTION-REPLY


    typedef G85501ASN::SNTPListType
            subnetworkTerminationPointsConnectedType;
    //ACTION-REPLY


    typedef M3100ASN::UserLabelType userLabelInUseType;
    //ACTION-REPLY


    typedef M3100ASN::DirectionalityType
            wrongDirectionalityType;
    //ACTION-REPLY


    // simpleSubnetworkConnectionPerformer
        interface SimpleSubnetworkConnectionPerformer:X721::Top {
    // simpleSubnetworkConnectionPerformerPackage
    // This managed object class is the engineering specification of the computational
    // interface simpleSncPerformerIfce
    // ACTION: setupSnc


    G85501ASN::SetupSncResultType setupSnc (
        in G85501ASN::SetupSncInformationType actionInfo)
        raises (ACTION_ERRORS, UsingMR)


    // this action is used to set up a Subnetwork Connection between subnetwork
    // termination points.
```

/ The behavior of this action is specified in *Simple Subnetwork Connection*
// *Configuration,* clause A.2.1 <G.854-01:OPERATION, setupSnc>

// PARAMETERS invalidTransportServiceCharacteristics,
// incorrectSubnetworkTerminationPoints,
//subnetworkTerminationPointsConnected, failure, wrongDirectionality,
// userLabelInUse ;
}

List of Acronyms and Abbreviations

ADM	Add/drop multiplexers
ADSL	Asymmetric digital subscriber line
AI	Adapted information
AIS	Alarm indication signal
AP	Access point
APS	Automatic protection switch
ASN.1	Abstract syntax notation 1
ASP	Access session provider
ASR	Access session requester
ATM	Asynchronous transfer mode
BEO	Basic engineering objects
BIP-n	Bit interleaved parity (of order) n
B-ISDN	Broadband integrated services digital network
BML	Business management layer
CI	Characteristic information
CORBA	Common Object Request Broker Architecture
CM	Connection matrix
CMIS	Common management information service
CMIP	Common management information protocol
CP	Connection point
CSP	Connectivity service provider
CSR	Connectivity service requester
CTP	Connection termination point

DCN	Data communications network
$\{D_0, R_0\}$	An object reference R_0 from domain D_0
DMI	Definition of management information
DXC	Digital cross-connect
EMF	Element management function
EML	Element management layer
E-R	Entity relationship
ESIOP	Environment-specific interoperability protocol
ETSI	European Telecommunications Standards Institute
FCAPS	Fault, configuration, accounting, performance, security
GDMO	Guidelines for the definition of managed objects
GII	Global information infrastructure
GIOP	Generic inter-ORB protocol
HO	Higher order
HOP	Higher order path
IAP	Internet access provider
IDL	Interface definition language
IETF	Internet Engineering Task Force
IIMC	ISO-Internet Management Coexistence group
IIOP	Internet inter-ORB protocol
IN	Intelligent networks
INAP	Intelligent Network Application Protocol
IOR	Interoperable object reference
IP	Internet protocol
ISO	International Organization for Standardization
IT	Information technology
ITU	International Telecommunications Union
JIDM	Joint Interdomain Management
LAN	Local area network
LC	Link connection
LEC	Local exchange carrier
LLA	Logical layered architecture
LT	Link termination
LTE	Line termination equipment
MAF	Management application function
MC	Matrix connection
MF	Mediation function

MI	Management information
MIB	Management information base
MIM	Management information model
MO	Managed object
MPEG	Motion Pictures Expert Group
NC	Network connection
NE	Network element
NEF	Network element function
NMF	Network Management Forum
NML	Network management layer
NNI	Network node interface
NNI-A	Network-to-network interface type A
NNI-B	Network-to-network interface type B
OA&M	Operations, administration and maintenance
OCL	Object constraint language
ODP	Open distributed processing
OID	Object identifier
OMG	Object Management Group
OO	Object-oriented
ORB	Object request broker
OS	Operations system
OSF	Operation systems function
OSI	Open systems interconnection
OTN	Optical transport network
PAI	Premise attachment interfaces
PC	Personal computer
PDH	Plesiochronous digital hierarchy
PDU	Protocol data unit
PRC	Primary reference clock
PSTN	Public switched telephony network
QAF	Q adapter function
QoS	Quality of service
RDI	Remote defect indicator
RDN	Relative distinguished name
RI	Remote information
RM-ODP	Reference model for open distributed processing
SCP	Service control point

SDH	Synchronous digital hierarchy
SLA	Service level agreement
SMI	Structure of management information
SML	Service management layer
SNC	Subnetwork connection
SNCP	Subnetwork connection protection
SNMP	Simple network management protocol
SOH	Section overhead
SONET	Synchronous optical network
SS7	Signaling system 7
SSF	Server signal failure
SSP	Service switch point
STM-N	Synchronous transport module (order) N
STS-N	Synchronous transport signal (order) N
TCP	Termination connection point
TDM	Time division multiplexed
TI	Timing information
TINA-C	Telecommunications Information Network Architecture Consortium
TMF	TeleManagement Forum (previously NMF)
TMN	Telecommunication management network
TSF	Trail signal failure
TTF	Trail termination function
TTP	Trail termination point
UDP	User datagram protocol
UML	Unified modeling language
USC	Usage session consumer
USP	Usage session provider
VC-n	Virtual container (order) N
VoD	Video on demand
VP	Virtual path
VPN	Virtual private network
WDM	Wavelength division multiplexed
WSF	Workstation function

Glossary

Access group
A group of colocated trail termination functions that are connected to the same subnetwork or link.
A group of colocated access points together with their associated trail termination functions.

Access point
A reference point where the output of an adaptation source function is bound to an input of a trail termination source or the output of a trail termination sink is bound to the input of an adaptation sink function. The access point is characterized by the adapted client layer characteristic information, which passes across it. A bidirectional access point is formed by an associated contradirectional pair.

Action
Used to express the service requests and associated responses that are exchanged between a client and the provider in executing a contract.

Activity
A set of ordered actions.

Adaptation
A transport processing function that consists of a colocated adaptation source and sink pair, which adapts a server layer to the needs of a client layer. The adaptation function defines the server/client association between the connection point and access point, and therefore these points delimit the adaptation function. Adaptation functions have been defined for many client/server interactions.

Adaptation function	An atomic function that passes a collection of information between layer networks by changing the way in which the collection of information is represented. An adaptation function is a transport processing function that presents the client layer network characteristic information at its output by processing the information presented at its input from the server layer network trail (sink). It accepts client layer network characteristic information at its input and processes it to allow transfer over a trail in the server layer network (source).
Adaptation sink	A transport processing function that presents the client layer network characteristic information at its output by processing the information presented at its input by the server layer network trail.
Adaptation source	A transport processing function that accepts client layer network characteristic information at its input and processes it to allow transfer over a trail (in the server layer network).
Agent	The part of a distributed application that manages the associated managed objects. It will respond to directives issued by a manager and transmit notifications reflecting the behavior and state of these objects.
Architectural component	Any item required to generically describe transport network functionality independent of implementation technology.
Atomic function	A function, which if divided into simpler functions would cease to be uniquely defined for transport networks. It is therefore indivisible from a network point of view.
Bidirectional connection	A connection consisting of an associated pair of unidirectional connections capable of simultaneously transferring information in opposite directions between their respective inputs and outputs.
Binding	A direct relationship between a transport processing function or transport entity and another transport processing function or transport entity that represents the static connectivity that cannot be directly modified by management action. Bindings represent the static connectivity within a network element. Binding relationships can never extend outside of network elements.

Broker	A broker puts consumers and providers in touch with each other.
Business contract	A contract where the client is a customer.
Business management layer	The part of the logical layered TMN architecture that has responsibility for the total enterprise.
Characteristic information	A signal of characteristic rate and format that is transferred within and between subnetworks and presented to an adaptation function for transport by the server layer network.
Client	A generic role within the service contract, which can be specialized according to its involvement in that service (e.g., user, caller, consumer, and subscriber).
Client layer	A G.805 network layer acting as a client with regard to transport services provided by a G.805 network server layer.
Client/server relationship	The association between layer networks that is performed by an adaptation function to allow the link connection in the client layer network to be supported by a trail in the server layer network. Client/server relationships never exist outside of network elements.
Community	An RM-ODP enterprise viewpoint concept representing a group of real-world entities bound together by an agreement to work together to achieve a particular purpose.
Compound function	A function that represents a collection of atomic functions within one layer. Example: A combination of several atomic adaptation functions within a certain layer (each serving one client layer) is a compound adaptation function. A combination of a (compound) adaptation function and the layer's termination function is a compound function.
Computational viewpoint	The RM-ODP viewpoint concerned with the functional decomposition of the system into a set of objects that interact at interfaces, enabling system distribution.
Connection	A transport entity that consists of an associated pair of unidirectional connections capable of simultaneously transferring information in opposite directions between their respective inputs and outputs. A connection defines

	the association between the connection points and the connection points delimit the connection.
Connection function	An atomic function within a layer that, if connectivity exists, relays a collection of items of information between groups of atomic functions. It does not modify the members of this collection of items of information, although it may terminate any switching protocol information and act upon it. Any connectivity restrictions between inputs and outputs are stated.
Connection point	A reference point where the output of a trail termination source or a connection is bound to the input of another connection or where the output of a connection is bound to the input of a trail termination sink or another connection. The connection point is characterized by the information that passes across it. A bidirectional connection point is formed by the association of a contradirectional pair.
Connection supervision	The process of monitoring the integrity of a connection or tandem connection that is part of a trail.
Connection termination point	A connection termination point resource encapsulates the management view of the G.805 adaptation function port and the part of this adaptation function that has a 1:1 relationship with the client signal. Thus, it represents the potential end of a link connection and the signal state at idealized points.
Contract	An agreement resulting from a negotiation between a client and a provider for furnishing a service.
Contract type	All possible contract clauses used as the basis of the negotiation between the provider and the client.
Customer	A client that renders payment for delivered services.
Distributed computing	The general concept of computer systems whose parts are spatially separated.
Domain	A grouping of entities, the members of which have some common defining characteristic, for example:

- An administrative domain in networking is usually defined as a collection of hosts, applications, and the interconnecting network(s), managed by a single administrative authority;
- An Internet "domain" is most commonly used to refer to a group of computers whose hostnames share a common suffix;

- An object computing domain is a scope in which a collection of objects, said to be members of the domain, are associated with some common characteristic.

Element management layer	The part of the TMN logical layered architecture that manages each network element on an individual or group basis and supports an abstraction of the functions provided by the network element layer.
Engineering viewpoint	The RM-ODP viewpoint concerned with the infrastructure required to support system distribution.
Enterprise viewpoint	The RM-ODP viewpoint concerned with the purpose, scope, and policies governing the activities of the specified system within the organization of which it is a part.
Equipment functional specification	A collection of atomic, compound, or major compound functions and any overall performance objectives that describe the functionality of equipment.
Fault	Refers to the inability of an item to perform a required function, excluding that inability due to preventive maintenance, lack of external resources, or planned actions.
Function	A process that acts on a collection of input information to produce a collection of output information. A function is distinguished by the way in which characteristics of the collection or of members of the collection of output information differ from characteristics of members of the collection of input information.
Global name space	A construct in which any object that is a member of that name space has the same name anywhere.
Global naming schema	A naming convention that is unique everywhere, where name spaces are created that are only global in the sense that any object anywhere can be named but not necessarily by the same name everywhere.
Information viewpoint	The RM-ODP viewpoint concerned with the kinds of information handled by the system and constraints on the value changes, use, and interpretation of that information.
Layer	A concept used to allow network functionality to be described hierarchically as successive levels; each layer being solely concerned with the generation and transfer of its characteristic information.
Layer network	A topological component that includes both transport entities and transport processing functions that describe

	the generation, transport, and termination of a particular characteristic information.
Link	A topological component that describes a fixed relationship between a subnetwork or access group and another subnetwork or access group.
Link connection	A transport entity that transfers information between ports across a link. A transport entity provided by the client/server association. It is formed by a near-end adaptation function, a server trail, and a far-end adaptation function between connection points. It can be configured as part of the trail management process in the associated server layer.
Managed objects	Conceptual views of resources that are either being managed or are there to support certain management functions.
Management point	A reference point where the output of an atomic function is bound to the input of the element management function or where the output of the element management function is bound to the input of an atomic function.
Manager	The part of a distributed application that issues management operation directives and receives notifications.
Matrix	A topological component used to effect routing and management. A matrix is contained within one physical node, and represents the limit to the recursive partitioning of a subnetwork.
Matrix connection	A subnetwork connection that is a connection across a matrix, formed by the association of ports on the boundary of the matrix. It may be configured as part of the trail management process or may be fixed.
Mediation function: TMN	May store, adapt, filter, threshold, and/or condense information passing between a management system and a network element or between management systems to assure information is in a form that can be mutually understood.
Model	A construct that corresponds to some aspect of reality in that the features shared in common by the model and that reality indicate what is deemed to be significant by the modeler. There may be several models of the same reality, in which different features are deemed to be significant.

Network	All of the entities (such as equipment, plant, and facilities) that together provide communication services.
Network connection	A transport entity formed by a series of contiguous link connections and/or subnetwork connections between termination connection points.
Network management layer	The part of the TMN logical layered architecture that has the responsibility for the management of a network as supported by the element management layer; for example, control and coordination of the network view of all network elements within its scope or domain.
Network node interface	The interface at a network node that is used to interconnect with another network node.
Object-oriented paradigm	Uses a simulation model of computation, encompassing the principles of: encapsulation, information hiding, message passing and late binding, class/instance/object, generalization, and relationships.
Open distributed processing framework of abstractions	Provides a means for separating the logical specification of required behaviors (i.e., functional requirements) from the specifications of physical architectures implemented to realize them.
Packages	Used to group object classes and their defined relationships and may be contained in other packages. They enable a complete and clear organization of a system specification as well as reuse of fragments of specifications.
Pairing	A relationship between sink and source transport processing functions or two contradirectional unidirectional transport entities or between unidirectional reference points that have been associated for the purposes of bidirectional transport.
Point-to-multipoint connection	A connection capable of transferring information from a single input to multiple outputs.
Policy	Specified as a set of applicable rules by either the client or by the provider in a community.
(G.805) Port	Represents the output of a trail termination source or unidirectional link connection and the input to a trail termination sink or unidirectional link connection. (For management purposes, a port is a role that can be played

	either by a connection termination point or a trail termination point.) It consists of a pair of unidirectional ports.
Premise attachment interface	An interface representing where access networks are connected with customer premises equipment or networks.
Reference point	An architectural component, which is formed by the binding between inputs and outputs of transport processing functions and/or transport entities. It is characterized by the information that passes across it and is the delimiter of a function.
Remote defect indicator	A signal that conveys the defect status of the characteristic information received by the trail termination sink function back to the network element, which contains the characteristic information originating trail termination source function.
Remote error indicator	A signal that conveys either the exact or truncated number of error detection code violations within the characteristic information (as detected by the trail termination sink function) back to the network element that contains the characteristic information originating trail termination source function.
Remote information	Information flow from sink direction to source direction of the same atomic function in unidirectional representation, containing information to be transported to the remote end, such as RDI and REI.
Retailer (TINA-C)	A retailer sells services on behalf of others (which may be other retailers or third party service providers).
Role	An RM-ODP enterprise viewpoint concept where each role corresponds to a specific community entity responsibility in providing a service to a client.
Routing	The process whereby a num)er of connection functions within the same layer are configured to provide a trail between defined termination points.
Server layer	The layer providing transport to a client G.805 network layer as part of a client/server relationship.
Service level agreement	Another way of referring to a business contract.
Service management layer	The part of the TMN logical layered architecture that is concerned with and responsible for the contractual aspects of services that are being provided to customers or available to potential new customers.

Subnetwork	A topological component used to effect routing of a specific characteristic information. It describes the potential for subnetwork connections across the subnetwork and can be partitioned into interconnected subnetworks and links. Each subnetwork in turn can be partitioned into smaller subnetworks and links and so on. A subnetwork may be contained within one physical node.
Subnetwork connection	A transport entity that transfers information across a subnetwork and is formed by the association of ports on the boundary of the subnetwork. It can be configured as part of the trail management process.
Systems engineering	The effective application of scientific and engineering efforts to transform an operational need into a defined system configuration through the top-down iterative process of requirements definition, functional analysis, allocation, synthesis, design optimization, testing, and evaluation. It integrates related technical parameters and ensures compatibility of all physical, functional, and program interfaces in a manner that optimizes the total system definition and design. It integrates reliability, maintainability, safety, and other such factors into the total engineering effort to meet cost, schedule, and technical performance objectives.
Tandem connection	An arbitrary series of contiguous link connections and/or subnetwork connections.
Technology viewpoint	The RM-ODP viewpoint concerned with the choice of technology to support system distribution.
Termination connection point	A reference point that consists of a pair of colocated unidirectional termination connection points and therefore represents the binding of a trail termination to a bidirectional connection. A special case of a connection point is where a trail termination function is bound to an adaptation function or a matrix.
Timing point	A reference point where an output of the synchronization distribution layer is bound to the input of an adaptation source or connection function or where the output of an adaptation sink function is bound to an input of the synchronization distribution layer.
Topological component	An architectural component used to describe the transport network in terms of the topological relationships between sets of points within the same layer network.

	A topological description in terms of these components describes the routing possibilities of the network and hence its ability to support transport entities.
Trail	A transport entity in a server layer responsible for the integrity of transfer of characteristic information from one or more client network layers between server layer access points. It defines the association between access points in the same transport network layer. It is formed by combining a near-end trail termination function, a network connection and a far-end trail termination function.
Trail management process	Configuration of network resources during network operation for the purposes of allocation, re-allocation, and routing of trails to provide transport to client networks.
Trail termination	A transport processing function that generates the characteristic information of a layer network and ensures integrity of that characteristic information. It consists of a colocated trail termination source and sink pair. The trail termination defines the association between the access point and termination connection point, and these points therefore delimit the trail termination.
Trail termination function	An atomic function within a layer that generates, adds, and monitors information concerning the integrity and supervision of adapted information.
Trail termination point	The trail termination point resource encapsulates the management view of the G.805 termination function port and presents the information derived from the termination function and any adaptation function information that has a 1:1 relationship with the server layer.
Trail termination sink	A transport processing function that accepts the characteristic information of the .ayer network at its input, removes the information related to trail monitoring, and presents the remaining information at its output. Thus, it terminates a trail, extracts the trail overhead information, checks validity, and passes the adapted client layer network characteristic information to the adaptation function.
Trail termination source	A transport processing function that accepts adapted characteristic information from a client layer network at its input, adds information to allow the trail to be

	monitored, and presents the characteristic information of the layer network at its output. The trail termination source can operate without an input from a client layer network.
Transmission	The physical process of propagating information signals through a physical medium.
Transport	The functional process of transferring information between different locations.
Transport entity	An architectural component that transfers information between its inputs and outputs within a layer network. Transport entities may be bound to each other or to transport processing functions; the points at which they are bound are the reference points of the transport network.
Transport network	The functional resources of the network that convey user information between locations.
Transport network layer	A topological component solely concerned with the generation and transfer of particular characteristic information.
Transport processing function	An architectural component defined by the information processing that is performed between its inputs and outputs. Either the input or output must be inside a layer network; the corresponding output or input may be in the management network (e.g., output of a monitor function). It has one or more inputs and one or more outputs that may be associated with inputs and outputs of other functions and entities. Such associations are termed binding relationships.
Unidirectional access point	A reference point where the output of a trail termination sink is bound to the input of an adaptation sink or the output of an adaptation source function is bound to an input of a trail termination source.
Unidirectional connection	A transport entity/connection that transfers information transparently from input to output.
Unidirectional connection point	A reference point that represents the binding of the output of a unidirectional connection to the input of another unidirectional connection.
Unidirectional port	Represents the output of a trail termination source or unidirectional link connection, or the input to a trail termination sink or unidirectional link connection.

Unidirectional termination connection point	A reference point that represents the following bindings: output of a trail termination source to the input of a unidirectional connection or the output of a unidirectional connection to the input of a trail termination sink.
Unidirecttional trail	A transport entity responsible for the transfer of information from the input of a trail termination source to the output of a trail termination sink. The integrity of the information transfer is monitored. It is formed by combining trail termination functions and a network connection.
Use case	In UML the functional requirements of a system are captured from the point of view of roles (or actors) that make some usage of the system. These roles and their relationship to types of usage of the system are analyzed one at a time, and each type of usage is called a *use case*.
Viewpoint	An abstraction that yields a specification of the whole system as related to a particular set of concerns.
Workstation function	Provides the means to interpret TMN information for the human user and vice versa.

About the Authors

Eve L. Varma holds B.A. and M.A. degrees in physics from the City University of New York. She is currently a manager of an advanced networking architecture and technology group at Lucent Technologies, Holmdel, New Jersey, where her primary work responsibilities include optical data networking architecture and management, support for development of optical transport network architecture and management specifications, assessment of software/distributed computing technology, and global standards development. In previous assignments, she worked on development of transmission jitter requirements, systems engineering and standards for transport and network management applications, and associated enabling technology and methodology development. She also led the Lucent team responsible for designing and prototyping a distributed optical network management system as part of the Multiwavelength Optical Networking (MONET) Consortium. Ms. Varma is a member of the IEEE and co-author of *Jitter in Digital Transmission Systems* (published by Artech House).

Thierry Stephant is a graduate of the Telecommunication National Institute (INT, Paris). In France Telecom Research Center (CNET), he is head of a team responsible for studies on TMN. Applications started with SDH network element management in 1990, then activities have been extended to allow for TMN specifications in a distributed environment with an application to SDH and ATM network management. More recently, the research activities have been oriented toward convergence between control and management in support of distributed infrastructure (such as CORBA) to increase network and service efficiency with an application to IP technology and its applications.

Antonio Rodríguez-Moral holds an M.S. in electrical engineering from the Technical University of Madrid, Spain. He is currently a member of the technical staff at Bell Laboratories, Lucent Technologies in Holmdel, New Jersey. His primary work responsibilities include research and architecture of optical data networks and next generation packet networks. In previous assignments, he worked on distributed network management, real-time object systems, system engineering and standards for Telecommunication Management Networks, and the development of SDH and Fiber-to-the-Home systems. He contributed to ITU-T and OMG to the definition and architecture of distributed object technologies and their application to real-time and network management systems. Mr. Rodríguez-Moral is a member of the IEEE.

Christine Pageot-Millet has a Ph.D. in computer science from Rennes University in France. She primarily worked in the field of artificial intelligence in the France Telecom Research Center (CNET, or Centre National d'Etudes des Télécommunications) in Lannion. She then joined the network management team where she contributed to the modeling and specification of network management services. This work has led to the definition of a modeling methodology based on the use of open distributed techniques, resulting in an ITU standard. She participated in ETSI and ITU meetings as contributor and editor of a standard document defining the vocabulary and concepts used to describe transport networks in order to manage them. She is now in a new team in charge of the definition and specification of end-user telecommunication services.

George W. Newsome holds a B.Sc. degree in electrical engineering from University College, London. He is currently a Distinguished Member of Technical Staff in the Advanced Networking Technologies group at Lucent Technologies, Holmdel, where his primary work currently includes optical networking architectures and global standards development. In previous assignments, Mr. Newsome has both created and managed the development of Network Element software, and has been involved with functional modeling since its inception in the ITU in the 1988–1992 study period. He has also worked on information modeling for network management and continued that work into the current ODP-based approaches. Mr. Newsome is a Chartered Engineer, Member of the IEE, and Senior Member of the IEEE.

Dennis K. Doherty holds an M.S.E. degree in computer science from Princeton University. He is currently heading up Lucent Technologies' product management team for transport network control and management systems in Holmdel, New Jersey. He has held a variety of positions in previous assignments related to the field of network management including market development in Europe and Asia, global standards for network management applications and technologies, systems, architecture and applications engineering, and project management. Mr. Doherty is a member of the IEEE.

Jean-Michel Cornily holds a Ph.D. in computer science from the Université de Rennes, France. He is currently a telecom engineer at CNET at France Telecom, Lannion, France. He is involved in telecommunication object-oriented and distributed software design. In previous assignments, he worked on transport network network-level information modeling. He participated in the definition of a new specification methodology for network management applications and in the evaluation of object-oriented techniques for specifying broadband switching software. Mr. Cornily contributed to ITU-T and ETSI and is involved in the OMG Analysis and Design Task Force activities.

Index

Abstract Syntax Notation (ASN.1), 39, 41, 129, 130, 169
Access point (AP), 68
Adaptation functions, 66
Adapted information (AI), 66
Add/drop multiplexers (ADM), 78, 81
Administrative domain, 237
Aggregation, 213
Alarms, 89
Allomorphism, 21
Announcement operations, 122
Anomalies, 89
Application planning. *See* Planning
Associations, 212
Asymmetric digital subscriber line (ADSL), 269, 275
Asynchronous transfer mode (ATM), 1, 43, 55
    layer, 191
    link, 191
    management services, 149
    multitechnology example, 281–84
    virtual path network example, 192
Asynchrony, 23
Atomic functions, 67, 69, 84, 85, 86, 198
Autonomy, 23

Basic engineering objects (BEO), 131–32
Behavioral compatibility, 20
Binding, 68
    explicit, 121
    implicit, 121
    late, 20
    name, 242
Booch approach, 209
Business management layer (BML), 36

Centralized work centers, 17
Characteristic information (CI), 59, 196
Circuit-switched networks, 81, 84
Class/instance/object, 20
Client
    alignment, 66
    encoding, 66
    labeling, 66
    logical, signal layer, 59
Collaboration diagrams, 215
Common management information protocol (CMIP), 31, 129, 246
Common management information service (CMIS), 31
Common management information service element (CMISE), 33
Common Object Request Broker Architecture (CORBA) standards, 24, 129
Community, 109–10, 111
Computational interface, 119
Computational objects, 118
Computational viewpoint, 220
    actions, 118
    announcement operations, 122
    child interface type, 121
    children, 120

Recent Titles in the Artech House Telecommunications Library

Vinton G. Cerf, Senior Series Editor

Understanding Modern Telecommunications and the Information Superhighway, John G. Nellist and Elliott M. Gilbert

Understanding Networking Technology: Concepts, Terms, and Trends, Second Edition, Mark Norris

Understanding Token Ring: Protocols and Standards, James T. Carlo, Robert D. Love, Michael S. Siegel, and Kenneth T. Wilson

Videoconferencing and Videotelephony: Technology and Standards, Second Edition, Richard Schaphorst

Visual Telephony, Edward A. Daly and Kathleen J. Hansell

World-Class Telecommunications Service Development, Ellen P. Ward

For further information on these and other Artech House titles, including previously considered out-of-print books now available through our In-Print-Forever® (IPF®) program, contact:

Artech House	Artech House
685 Canton Street	46 Gillingham Street
Norwood, MA 02062	London SW1V 1AH UK
Phone: 781-769-9750	Phone: +44 (0)20 7596-8750
Fax: 781-769-6334	Fax: +44 (0)20 7630-0166
e-mail: artech@artechhouse.com	e-mail: artech-uk@artechhouse.com

Find us on the World Wide Web at:
www.artechhouse.com

DATE DUE

APR 26 2001